BEYOND

By the same author:

Noise Pollution
London's Drowning
Floodshock
Our Drowning World
Earth's Changing Climate
The Fate of the Dinosaurs

BEYOND

THE WARMING

The Hazards of
Climate Prediction
in the Age of Chaos

Antony Milne

PRISM
PRESS

First published in Great Britain 1996 by
PRISM PRESS
The Thatched Cottage
Partway Lane
Hazelbury Bryan
Sturminster Newton
Dorset DT10 2DP

and distributed in the USA by
ATRIUM PUBLISHERS GROUP
3356 COFFEY LANE
SANTA ROSA
CA 95403

ISBN 1 85327 098 9

Typeset by Avonset, Midsomer Norton, Bath
Printed by The Guernsey Press Ltd., C.I.

CONTENTS

INTRODUCTION

This book aims to bring readers up to date with the subject of climate change as it appears in the mid 1990s, and to try to counterbalance the rather heavy anthropocentric bias evident in earlier popular and general works on climatology. It tries to put the greenhouse effect into context by focusing in depth on the postwar climate record, and by placing a greater emphasis on the fascinating complexities of the earth sciences, although it assumes no prior knowledge of the subjects by the reader.

The central point of Chapter One is that the Earth, with 70 per cent of its surface covered by water, cools itself by surface evaporation, and that atmospheric gases (apart from water vapour) play only a minor role in determining climatic change because they function by radiative, rather than convective, processes.

In Chapter Two I examine the role that the carbon cycle has played in Earth's geophysical history, and challenge the accepted wisdom that carbon dioxide will inevitably accumulate in the atmosphere because all of Earths' carbon reservoirs (or 'sinks') are now full to capacity. Indeed, I point out that many scientists are puzzled to note that much of the Earth's stock of carbon has actually 'gone missing'. In prehistoric times there were often astonishingly high levels of CO_2, although such levels seem to have had a poor correlation with Earth temperatures.

In Chapter Three I examine the way that other pollutants in the atmosphere, especially the sulphates, may be bringing about a cooling in many industrialized regions of the world. I then turn in Chapter Four to a study of the oceans which regulate and dominate the world's climate. I discuss the various global ocean monitoring experiments that are currently being undertaken with the aid of satellite surveillance craft to help find out whether the oceans are indeed warming, and highlight the technical difficulties in doing this.

Chapter Five deals largely with the El Nino effect and Earth's frequent climatic 'flips', which seem to have something to do with reversing warm and cold ocean currents. These reversals also seem to be related to Earth's periodic ice ages, and I point to new evidence that shows how surprisingly rapid these events have been in Earth's history.

The following chapter deals with the scientific aspects of understanding climatic change, and explains in some detail the instrumental and computer aspects of the subject, and the way 'General Circulation Models' are created and used. Emphasis is put on the way 'chaos theory' can so easily undermine the predictive potential that climatologists seek to acquire.

Lastly but by no means least the role of the sun as the prime mover of weather systems is dealt with. Space is given to the views of a growing number of scientists who believe that the sunspot cycle, and its interaction with the magnetic features of atmospheric gases and weather dynamics, may have something to do with the somewhat inexplicable changes in the historic climate.

This book, then, is about science and scientific methods. It should become clear that our present understanding of the climate is derived from a great web of information gathered from many hi-tech sources from around the globe. Weather-information collecting and processing, using extremely expensive computers, as well as complex undertakings such as ice-core drillings and oceanic satellite surveillance, employ a vast number of scientists and engineers in many countries; almost as many, in fact, as do some military or space-flight organizations. It is in fact a space-age business, and the scope of the various international organizations now monitoring the Earth's geosphere and the Sun's energy sources with the aid of spacecraft has grown considerably, and its participants are engaged in the same sort of explorative adventure. As we head for the new millenium climate scientists are now indeed going Beyond the Warming.

And in just a few years from now a giant coloured visual display terminal may bleep a hazard warning: 'Climate cooling within ten years confirmed!'

Chapter One:

IS THERE A WARMING?

The new climate report on the greenhouse effect was important. Published in 1990 it was deemed to have set the climate agenda, and has succeeded in so doing. It was sponsored by the United Nations (UN) and its World Meteorological Organization (WMO), and the UN Environmental Programme (UNEP), representing the findings of the Inter-governmental Panel on Climate Change (IPCC). It was launched by Sir John Houghton, then Chief Executive of the British Meteorological Office.[1] Thus promoted and funded by four powerful global organizations, three of them esteemed scientific establishments, it could hardly be ignored by the world community. If there was any one definitive treatise or thesis that dealt unequivocally and authoritatively with the greenhouse effect, it had to be the IPCC report.

It was therefore surprising to have not only its findings and predictions scathingly challenged as soon as it was published, but also its methodology and science. Its principal critic, Richard Lindzen, Alfred P. Sloan Professor of Meteorology at the Massachussetts Institute of Technology (MIT), said that the strangeness of the document became apparent on the very first page, with declarations reminding us that its findings had been recorded and interpreted with the utmost rigour. No true body of scientists, maintained Lindzen, would ever speak with such confidence in dealing with such an immensely complicated and poorly understood subject as climate change.[2]

He pointed out that it was semi-popular, written in the style of *Scientific American*. Others said it looked like a large, floppy car manual. It had many graphs, tables, sub-headings and formulae, and yet contained only two mathematical equations. It was too obviously a 'translation', as the preface-writers put it, of something else.

Lindzen suggested this was because pressure had been put on the IPCC scientists to come up with a doom-laden thesis, even to the extent of biasing a report that was probably, in its original form, more balanced and evaluative.

One important statement in the report went to the heart of the issue. It said: 'Because of the strong theoretical basis for enhanced greenhouse warming, there is considerable concern about the potential climatic effects ....'[3] This disturbed Lindzen precisely because observers should never be impressed with the 'strength' of any theory; rather they ought to be more impressed with the *validation* of it.

The 'theory' itself was based on nothing less than the modest scientific fact, repeated time and again by science journalists, that there are natural greenhouse gases present in the atmosphere, such as carbon dioxide, which make the Earth warmer than it would otherwise be. As carbon dioxide is observed to be increasing then, ergo, the Earth must be getting warmer, because CO^2 traps incoming solar radiation that would otherwise escape back into space.

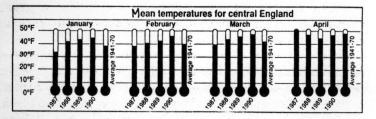

No statistical warming: Although English winter temperatures appeared to be rising during the 1980's the periodic cold snap, such as occurred in the winter of 1987, and April 1989, returns the average to that prevailing from 1941 to 1970.

Source: Daily Telegraph

However, the argument is always framed within an astronomical perspective which, rather than supporting the theory, should actually ring theoretical alarm bells. There is, for example, the obligatory reference to Mars and Venus. 'From this point on', wrote Lindzen, 'the summary is a downhill ride' because such a simplified explanation 'has a seductive inevitability about it'. In other words, from false analogies, false assumptions are made. Mars, despite having 20 times more carbon dioxide than Earth, is a relatively cold planet. Venus, of course, with 90 per cent of its atmosphere in the form of CO_2, is indeed very hot, and yet these high temperatures are not entirely due to the radiative properties of carbon dioxide.[4] If they were, Earth, with 0.03 per cent of its atmosphere consisting of carbon dioxide, would be considerably warmer than it is now even though the percentage, by itself, seems so tiny. This is because Earth does not cool primarily by radiation but by surface evaporation, which Venus does not experience. After all, a true backyard greenhouse on Earth is warm inside not just because it traps the Sun's radiation through panes of glass, but because there are no surface winds inside the greenhouse to draw off any of the heat.

But Earth's surface heat *is* carried away by air turbulence and meteorological systems to be deposited at higher altitudes and latitudes, sometimes in the form of snow and ice packs which themselves can have quite unexpected effects on temperature gradients (see Chapter Five). But if a planet has no Earth-like combination of land and oceans, and ice, it has no climate; if it has no climate it has no changing monthly and hardly any seasonal temperatures, hence it will be impossible for surface gases alone to determine changing temperatures, except over *millions* of years. Climatologists, when referring to 'global warming', are in danger of making un-climatological statements.

Earth's temperatures, in other words, are governed by physics rather than chemistry. Understanding the role of water, moisture or clouds is vital. Any planet that has 70 per cent of its surface covered in depth by water will *never* be able to achieve very high atmospheric temperatures. Indeed, away from the equatorial zones of Earth and depending on the distance between land surfaces and surrounding oceanic currents, the main annoyance to warm-blooded creatures like us will come from temperatures that are frequently too cool. Naturally, if Earth were suddenly thrust closer to the Sun, say because it had swapped orbits with Venus, then the dramatic rise in temperatures would simply boil away all the oceans. Then, becoming a rocky planet, Earth would soon acquire a dense carbon dioxide atmosphere as the carbon-rich rocks baked under a relentless Sun unobscured by clouds. Until this unlikely event happens, and because water exists only in the liquid state at between 0 and 30C, we are protected from a runaway greenhouse effect even if CO_2 levels are 10 times what they are now (i.e. if they go from 0.03 per cent of the total to 3 per cent of the total of atmospheric gases). Even if CO_2

were one hundred times as great, the Earth would still not warm excessively. All animal life would suffocate, and forests would grow as high as skyscrapers and fierce hurricanes would rage continuously, but Earth would not go the way of Venus.

The Earth is a wet planet; what rescues it from excessive warming is something known as *latent heat transfer*. Present computer models do take into account latent heat transfer, but poorly so. Computers cannot even predict existing temperatures without some 'tuning'; hardly a sign of precision science. Nor do climatologists mention the need to improve transfer (or 'transport') modelling, despite the vital importance of this for assessing the Earth's present temperatures, or calculating its future temperature. Indeed one IPCC chapter goes to considerable lengths to disguise the importance of water vapour feedbacks in the scenario. The IPCC report says that CO^2 accounts for 25 per cent of the upper atmosphere warming, while water vapour and clouds account for some 65 per cent. Lindzen puts the former figure much lower, and elevates water vapour to 97 per cent, especially the small proportion that is above six kilometres, where greenhouse gases are much thinner (see Chapter Three).

Nor is there any evidence, as claimed by pro-greenhouse advocates, of sea level rises. This is a subject in any event complicated by isostacy — the rise and fall of sea levels in relationship to similar movements in the land, and the pattern across the globe is not consistent. David Aubrey, formerly head of Coastal Research at the Woods Hole Oceanographic Institute in America, one of the most influential of its kind, pointed out a few years ago that: 'You can come up with whatever answer you want, in effect. No useful conclusions about the future (of sea level rises) can be drawn from the record. There is no evidence that the sea level is rising due to global warming'.[5] Submarine data hinting that ice thickness was lessening were also suspect because they often came from 'snapshots' taken 10 years apart, and were contradicted by satellite observations.[6]

The idea that sea level changes may have something to do with the changing shape of the ocean floor which may in turn have something to do with plate tectonics is put forward by geologists Lawrence Cathless of Cornell University and Anthony Hallam of Birmingham. They suggest the mooted rises of many tens of feet would necessitate melting half the Antarctic ice. Even extinctions of marine life due to sea level changes might be caused by magma rushing up through a rift in the seabed, which would bind with dissolved oxygen, stripping it from the water and causing widespread suffocation.[7]

Richard Lindzen also told a meeting of the Royal Meteorological Society in London in December 1992 that global temperatures are very unresponsive to changes in overall radiation from whatever source. Most feedbacks, such as the lessening of albedo as the ice packs shrink, and the increasing cloud cover as surface heat builds up, and the microbiological reaction to changes in atmospheric temperature and chemistry, are *negative*,[8] i.e. self-adjusting.

Do the Data Prove a Warming?

The other aspect of the debate concerns the historic climate. There has, we are told, been a measurable warming in the past 100 years. Climatologists at the Climate Research Unit at the University of East Anglia say that 'the causes of the warming are less certain than the trend itself',[9] which they put at between 0.8C and 2.6C.[10] The IPCC report talks of a 'real, but irregular' warming, and confirms, in its most recent 1993 report, earlier estimates of a likely 2-4C warming over the next 100 years.[11] The truth is that 100 years is an eyeblink in comparison with the age of the Earth, in particular the hundreds of millions of years that have passed since Pangea first broke up and for the first time brought climates into existence.

The mooted rise in temperatures over the past century cannot actually be detected in the US where observations are numerous and accurate.[12] Scientists at the American Weather Bureau have produced a study showing that there is no evidence the US has warmed or chilled significantly over the last century.[13] Dr Sherwood Idso, at the US Water Conservation Laboratory, who has studied CO_2 for the past 25 years, said ½C warming over the past 100 years is 'tenuous'and not apparent in the 'real world'.[14]

Keith Shine of the University of Reading says there was a 0.2C warming during the 1980s, but this is surely within a natural range of error, and would in any event be difficult to attribute to a carbon dioxide warming. 'Hence', he says, 'many scientists use the double negative "not inconsistent" to describe the link between observed temperature trends and changes in CO_2'. In other words they say it 'might be'.[15]

Further, the George C Marshall Institute, a privately funded and highly influential foundation based in Washington, and a highly influential group responsible, in part, for shaping President Bush's dissenting views on the greenhouse effect at the time international conferences were debating the subject, said they saw no evidence of a recent anthropogenic greenhouse effect, putting the blame on solar factors (see Chapter Seven). The Institute pointed out that there should be a greater rise than 0.4C during the 20th century if anthropogenic warming were to blame.[16] Jerome Namias of the Scripps Oceanic Institute, a long-range weather expert, was among others who wrote a letter to George Bush urging that no action be taken to curb greenhouse-gas emissions.

A prestigious critic of the greenhouse effect is William Nierenberg, a former director emeritus at the Scripps Institution of Oceanography at La Jolla, California. He co-authored the report that expressed the dissenting views of the George C Marshall Institute, and didn't so much criticise the greenhouse message as the climatic exaggerations made by its messengers.[17] He criticised computer models which are based on highly theoretical calculations concerning predicted climatic extremes that gave results 'that are inconsistent with the record of the last 100 years [and show] serious inconsistencies in estimates of recent global temperature changes'.[19] Most commentators point out that most of the greenhouse gases are said to have been emitted in postwar years, the very period that has been characterised by lack of consistency in meteorological trends.[20]

In fact a more detailed breakdown of the 100-year data reveals some curious anomalies. Historic climate data are derived not so much from sophisticated atmospheric gas or ice-core samples (see Chapter Five) but from rough averaging-out of temperature statistics collected by sundry researchers and meteorologists, professional and amateur, on the assumption that 10-year averages could be a reliable guide to longer-term trends.[20] Yet some talk of a 'prolonged cooling' from the 1920s to the 1960s.[21] Michael Allaby, an ecologist who was closely associated with the *Limits to Growth* publication of 1972, says there was a cooling from 1940 until about 1970, 'most strongly marked in the Arctic'.[22]

Other commentators point only to a postwar drop in temperatures in the western world. Since then the warming has 'resumed'. Allaby suggests that this cooling dragged down the average to make it plus 0.5C, implying it could otherwise have been higher. This, surely, is perverse logic, since the argument (based on statistical averages) could easily be reversed and the postwar argument of a general cooling spoiled because of several warm years intervening. Jim Angell, of the National Oceanic and Atmospheric Agency (NOAA), in America found a net warming of 0.24C only in the past 30 years,[23] a figure that can be regarded as virtually meaningless.

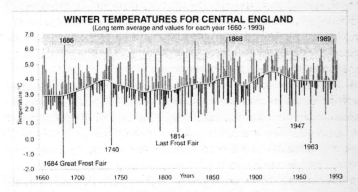

During the past 300 years, data shows there may have been a half a degree rise in temperatures. More significantly since the last war, during which time CO_2 emissions are said to have dramatically increased, winter temperatures have actually flattened out.

Source: Dr William Burroughs

There is much evidence of global temperatures rising long before the last century. We know from the historical record that the Danes settled in Greenland and Vikings sailed the North Atlantic during warm periods around 1,000 AD. Hubert Lamb, a former emeritus professor at the University of East Anglia's Climatology Unit (CRU), has spent most of his professional life piecing together documentary and historical evidence of climatic change. He says, quoting from various annals and records of harvests, there was a medieval warming. Physical sources such as pollen records and lake sediments also prove this.[24] Then a significant change caused the collapse of the Danish settlements, and severely inhibited further exploration of North America[25]. Other dramatic variations have revealed themselves. For example, though the second half of the 12th century was very warm, the first half was very cold. Eleventh and 13th-century summers were near normal. This was followed by a mini ice age from 1400 to 1800, which is generally known as the Little Ice Age (LIA). Nine degrees F is believed to be the difference separating the end of the last Ice Age about 12,000 years ago and the present.[26]

The tree-ring evidence (from which scientists can determine such factors as changes in atmospheric temperature, moisture and chemistry) points to a mini-chill which lasted from 1570 to 1650.

The term 'greenhouse effect' is a human invention. It is a great weakness on the part of the British in particular to think that strange weather is the prerogative of our times only. In every century without fail it is the lot of the English to learn through their own experience or through their newspapers how their country has been battered by intense storms and freak abnormal weather fronts, with questions asked in the House of Commons. To suppose otherwise is to indulge in temporal egoism. 'Our ignorance of history makes us slaves to our own times' said Gustave Flaubert. The Scots adopted a slave mentality, as did Robert Burns, when they took the union of the English and Scottish parliaments in 1707 to be an act of treachery; but it is likely the Scots were virtually starved into surrendering their independence by a bitter cold spell that played havoc with the harvests. Similarly, the Irish potato famine that between 1840 and 1850 either killed or forced into emigration three million Irish people, has been linked to a sharp rise in atmospheric humidity that spread potato blight across the land.

Even more recent British data prove little. Britain lies at the boundary of two weather systems: the North Atlantic 'maritime' system bringing cool, humid, overcast weather and rain, and the Eurasian 'continental' system which is mainly dry, hot and sunny. In some years the maritime system dominates over the other, and vice-versa. The outcome is determined largely by the jet streams that dislodge first Arctic air and then tropical air streams over Britain (see Chapter Five). This to-ing and fro-ing might explain much of Britain's erratic weather. The first half of the 18th century brought much more continental-type weather, with 50-day frosts in January and February 1709 and prolonged droughts in the 1740s.[27]

The long-term trend since 1700 does seem to have been for higher temperatures. But there have been other fluctuations, such as the run of hot, dry summers between 1772 and 1783 and another series of cool summers between 1809 and 1818. The interspersing of frigid with mild winters in Britain between 1979 and 1987 merely mirrored the fluctuations both on an historic and a global level, proving nothing about trends towards either a long-term warming or cooling.[28] What it does prove, however, is that subjective attitudes towards perceived trends in weather become more important when the objective scene is so inconclusive. For example Britain experienced 15 poor or average summers on the trot between 1950 and 1974 inclusive. This is generally admitted to have been a 'cool' period. However since then warm summers have only occurred in 1975, 1976, 1983, 1984, 1989, 1990 and 1994, i.e. seven years out of 21, hardly proof of a sustained greenhouse warming.

Taking the period 1950-1994 (44 years) we have a total of 29 average summers, so the figures are more likely to provide ammunition for cooling theorists.[29] Furthermore the year 1976 showed up *globally* as being exceptionally cold, although memories of most Britons is that it was one of the warmest years of the century, with 15 consecutive summer days recording temperatures of 90F.

Strangely, none of the data published in popular accounts of the greenhouse effect even remotely support the warming hypothesis. John Gribbin's 1990 *Hothouse Earth* is dominated by caveats, warnings and exceptions, with all the usual admissions about the difficulties involved in ascertaining how much the world has warmed.[30] He dismisses the findings of most research groups and concentrates solely, for the warming thesis, on data from the CRU and the Nasa institute in New York. Most of the warming, he says, seems to have been in the 1980s. Yet he dismisses a slight cooling in 1984 and 1985 which he attributes to El Chichon's volcanic eruption in 1982 as if this ought to be disregarded simply because it refutes the warming theory. In 1988 the CRU included new global data which, instead of proving a warming, actually brought the northern hemisphere average for 1981-4 down by 0.04C.[31] These, however, were 'temporary lulls' in an upward trend because one of the successes of the new CRU data was that more of the southern hemisphere was included for the first time, showing a more consistent 1980s warming.

Yet the global data still looks pretty paltry. After including new recording station information, and after eliminating urban biases, Mick Jones of the CRU 'looked in detail at the 20-year period from 1967 to 1986 . . . and found a total warming of 0.31C in the northern hemisphere and 0.23C in the southern hemisphere',[32] the latter figure being so low, and fraught with the hazards of marginal errors, as to lessen the impact of the 'consistent southern hemisphere warming' message. Gribbin admits the global warming was far from uniform, with Britain cooling by about 0.25C between the mid 1970s and the mid 1980s.[33] There was also a 'newly identified' cooling over Europe, eastern Canada, and southwest Greenland, with temperatures dropping over Scandinavia (which was unfortunately 'out of step' with its surroundings). Parts of the Antarctic coast had cooled even more dramatically by about 1C, while a warming had reached 2C over north western Canada, and 1.6C over western Siberia.

The Nasa data was just as dubious. Again it was mentioned because Nasa had more global coverage than other institutions, although we find that the northern latitudes warmed by 0.8C between the 1880s and 1940, cooled by 0.5C between 1940 and 1970 and had started to warm again in the 1970s.[34]

There are other published sources showing that meteorological records of rising temperatures in the past are not as reliable a guide as they first appear. James Elsner of Florida State University and Anastasios Tsonis of the University of Wisconsin described in 1991 a new analysis of three main data sets of northern hemispheric warming compiled during the 1980s by Soviet and US researchers. But projected nonlinear trends were not the same for the three data sets. According to Tsonis, 'The differences in the observed surface temperatures are the result of different populations rather than different samples from the same population'.[35] This means that none of the data sets can be used as a reliable guide as to how temperatures are likely to rise in the future until the reasons for the differences are understood.

And yet hot summers seem to stir the media into a frenzy, and scientists, perhaps against their better judgement, are caught up with the need to popularize doubtful explanations. The *Sunday Times* said on 24th July 1994 that 'scientists believe this year's warm weather, in which July temperatures are up to 4C hotter than normal' was something to do with Mount Pinatubo. Bill McGuire, a vulcanologist at University College, London, was quoted as saying that the eruption had indeed produced a 'global cooling' that had reduced temperatures by 0.4C. In other words the climate, having earlier gone into reverse, was now back to normal and was continuing with its warming trend. By saying this however McGuire merely confirmed that volcanoes played a very big role in determining how meteorological data would reveal themselves to researchers, even if they could not be proved to have a decisive influence on the historic climate.

Further, very little attempt is made to examine what is happening in the remaining seasons in order that global trends can be properly put into context. Extreme temperatures for short periods mean little. The onset of an Ice Age is determined more by a series of milder, but more snowy winters, with average or cool summers, so that the ice stays on the ground longer throughout the season, and the northern ice sheets gradually creep southwards as the years pass. Cool spring temperatures are the deciding feature in this scenario, and any later summer heat must be prolonged and continuous, year after year, to melt the ice in its tracks. One of the significant features of postwar years has been the evidence of long-drawn-out cool springs in the northern hemisphere in the last 25 years, with later hot spells being erratic or short-lived. Furthermore, as I pointed out in my book *Our Drowning World*, winters in both hemispheres were becoming dramatically unpredictable in ways associated with the past onset of Ice Ages. Mild winters are still interspersed with extremely frigid ones. Severe British winters have increased in frequency in the past 18 years compared with only five between 1940 and 1978.[36].

In any event scientific analysis of the Earth and its atmosphere reveals nothing out of the ordinary. Physicist Philip Abelson points out that satellite measurements prove only that wide variabilities in temperatures have taken place between 1979 and 1988[38]. North Pole atmospheric records show no signs of a warming even after 40 years of study.[38]

In as mathematically imprecise a science as meteorology the quality of the data depends on good maintenance of the instruments and the skills of the operator. Mick Jones and Tom Wigley of CRU say that computer models predict that changes in the radiation balance caused by greenhouse gases equals about 1 per cent of the luminoscity of the Sun, but even they admit the warming that has allegedly occurred reflects uncertainties due to lack of understanding of natural countervailing trends.[39] But Richard Lindzen says that greenhouse gases emit as much radiation as they absorb depending on their position in the atmosphere. The gases are somewhat interdependent upon each other.[40] Satellite measurements at high altitudes are very inadequate, and 'completely compromise model predictions'.[41] For example computer models point to temperature variations in different parts of Earth's gaseous envelope (such as cooling in the uppermost parts, and warming areas below this), with the greatest effect expected in the stratosphere. Scientists from the

Centre Nationale de Research Scientifique in Paris used a laser beam to point straight upwards into the upper atmosphere from an observatory.[42] Reflections bounced from the French beam off gas particles enabled the density and temperature of the atmosphere to be measured up to 62 miles above the ground. Yet the laser results were surprising. The mesosphere, about 40 miles above the Earth, has cooled by 5C since 1979, about twice as fast as most computer models have predicted.[43]

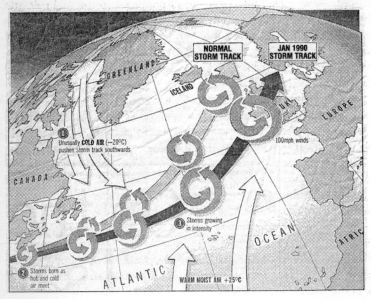

Many climatologists believe that virulent storms that occasionally wreak havoc in the northern hemisphere (as in this one that struck England in January 1990) are more likely to signal the onset of a new ice age than a warming.

Source: Sunday Times

Measuring the temperature of the Earth from space is immensely complicated. Satellites use sensitive devices known as microwave radiometers which scan the Earth's surface at an altitude of 10,000 feet and measure the amount of upwelling radiation. But it is a task that is not made easier by the presence of heat-emitting clouds. And virulent westerly winds in the far northern hemisphere can stir up low-lying cold air and make the temperature near the ground *seem* warmer to the sensors. In spite of this scientists from the Nasa

Marshall Space Flight Centre in Alabama have detected no temperature changes during a 15-year survey.[45] If anything, the temperature was trending *downwards* by 0.04 of a degree per decade.[46]

On the other hand the Arctic regions do seem to have warmed significantly since the War. Evidence from scientists at St Andrews University shows that flowers in the Arctic regions are growing at higher latitiudes than before, and trees in the Canadian North are 'poised' to move northwards. The problem is that vegetation has moved backwards and forwards throughout history. Professor Harvey Nichols, who has carried out extensive research in such regions for over 20 years, pointed out in March 1995 that the latitude beyond which trees did not grow was much further south some 20,000 years ago, but about 5,000 years ago it was 400 km further north than today. [47]

Considerable publicity was given to the giant iceberg that broke away from Antarctica in early 1995, 'roughly the size of Luxembourg',[48] and said by some to be definitive proof of a global warming. And yet throughout time similar icebergs have drifted away as snow accumulates and new ice squeezes out the old. In 1986 three 'megabergs' broke away from Antarctica, each bigger than the most recent. David Vaughan, a glaciologist, says that Antarctic temperatures since the 1940s have moved sharply up *and* down according to the amount of ice present.[49] Some oceanographers believe such events might precipitate a mini ice age by infusing the Southern Ocean and the Pacific with freezing, clear, water that can cause rapid current reversals that could in turn dramatically change inland weather patterns.

The Message of the Storms

The greenhouse effect is an impressive newcomer to the scene, aiming to take its place alongside the big league of 20th century revolutions in scientific understanding like relativity theory or continental drift. It has gained massive public cognizance and has been discussed endlessly in learned journals. But nowadays it seems to be punching above its weight, and remains a credible theory only by skillfully changing the verification ground rules in its favour.

No one was checking thermometers, but feature editors were noting something else which seemed to prove that a global warming was under way. Storms and hurricanes across large parts of the globe seemed to be becoming more frequent, virulent and damaging. It was even said that the disastrous losses at Lloyds, the world's most prestigious reinsurance company, were proof of global warming!

Some commentators suggest it all started with the Sahelian drought of 1968, which adversely affected the lifestyle of the nomads and their cattle,[43] and caused epidemics and intercommunal conflicts. Soon the strife-torn Sahel became an intellectual battlefield for competing schools of thought. It was widely believed by some that drought in the region was man-made, and environmental perspectives were beginning to loom larger. The 1970s were undoubtedly turbulent years. There were failures of monsoon rains in India, droughts and severe winters in Europe and America. The eighties were momentous. In 1986 the Deep South of America suffered a gruelling heatwave that was superimposed upon the worst drought in the south's history,[44] with thermometers rarely falling below the mid-90s throughout the summer.

Northern winters were seemingly also becoming unusually severe. In the winter of 1983/4 a blizzard in Denver crushed 14 buildings flat, and a chain of tornadoes claimed 80 lives in the eastern zones.[45] In Britain in May 1984 the Commercial Union insurance company hinted it would have to raise its premiums if future weather was to continue to be so awful.[46] A spokesman for the company said that during the last decade much greater extremes of weather had been experienced, and the earlier pattern of a bad winter every five or six years was being reduced to every two or three years.[47] General Accident complained that their underwriting accounts showed rising losses as a result of claims for weather damage. Senior Met men from the National Oceanic and Atmospheric Agency (NOAA) published an alarming scientific paper which showed that six out of eight consecutive winters from the mid 1970s to the early 80s were either excessively mild or extremely frigid, and such a combination would not be expected to occur for more than 1,000 years.[48]

A landmark event occurred in October 1987 in England, when a virulent storm raced across the south of the country, knocking down 19 million trees, 150,000 telephone lines and causing 19 deaths, and running up damage bills amounting to billions of pounds. It was said by some meteorologists to be a once-in-300 years event. The cause of the storm was said in an official Met Office enquiry report issued later to be due to the air above the sea being much warmer than usual so it picked up more evaporation energy.

But the Met Office computer was misled by an unusual set of weather circumstances. The storm came in unexpectedly from a more southerly position and veered much further northwards and towards England's coastline, to reach its maximum density over heavily populated areas. Many of the trees were still in full leaf presenting a larger surface area to the fierce winds. Trees were simply rocked back and forth until they keeled over. Yet the TV weather forecaster who had the previous evening dismissed a viewer's fears of anticipated hurricanes (something which the media refused to let him live down) highlighted the fact that Britain's stormy weather originates in the Atlantic where there are few permanent sources of information. In fact the number of weather ships in the eastern Atlantic has actually declined over the years.[49]

A year later the Greenhouse Effect was being discussed at cocktail parties and everyone, with a heightened sense of awareness, was cognizant of unpredictable deviations from the long-term weather average. An anomalous series of superlative 'supercyclones' appeared within a few months of each other. In 1988 there was Hurricane Gilbert, the 'strongest-ever' Caribbean hurricane. The following year gave birth to the 'most expensive' storm from the same area, Hugo, which struck South Carolina. Then there was Cyclone Ofa of 1991, causing damage of catastrophic proportions through much of the South Pacific. Less than a year later Cyclone Val, even more virulent, destroyed western Samoa, causing insurance companies to withdraw completely from those regions.[50] Hurricane Andrew in August 1992 was said to be one of the 'worst storms' this century.[51] For all these events to be merely random was, it was suggested, to strain statistical credibility to the limits.

However these were slippery times for science, with questionable modes of reasoning becoming evident, and too much attention being paid to the short-term. Some scientists fell into the trap of assuming the Greenhouse Effect already been to have proved, and that the weired links were a portent of things to come. The Sahel began to vary between normal and hot. For example the wet 1988 Sahel experience coincided with the hottest year ever.[52] Surprisingly, even over such a minuscule time frame, CRU scientists were saying the Sahel situation led them to 'come round to the view' that *the* warming was to blame.[53] Yet rainfall data in the region was hard to assess, as so few rain gauges are available. Rain was patchy, with fierce storms occurring in different parts of Ethiopia and Sudan. Yet scientists were convinced that rainfall for the entire region — Mauritania, Senegal, Mali and Sudan — were well down on the usual for 1990.[54]

Further confusion about the trend in climate came from a conference on drought in December 1991. Mike Hulme, a senior research associate at the CRU, had developed a global 'humidity index' that could form the basis for assessing the world's dryland areas. He estimated that between 1931 and 1990 as much as 63 per cent of the African continent had become drier. But large parts of Africa had also become cooler.[55] The index also revealed that the west of Australia, western Europe and the southern US were getting wetter, while it was getting drier in southern Africa, western South America and in the south-east of Asia. The causes — whether man-made or natural — are still not clear. Hulme said the dry African conditions could be the result of changes in sea surface temperatures (SSTs) (see Chapter Five).

As soon as the idea of greenhouse-heated seas entered the scenario, scientists began talking in apocalyptic terms. 'The disruption would be substantial', said John Topping, president of the Washington think tank, the Climate Institute. He predicted that over the next 80 years there would be an increase in storm damage.[56] Mercifully the worst of the altered-climate damage would have been done by the year 2010, and we could expect a return to more settled conditions by 2070. In the meantime rising sea levels and drastically altered weather patterns would affect the Indian subcontinent and the emerging Asian 'tiger' economies. Thousands of square miles of coastal areas would be at risk, displacing seven million people, and 10 per cent of Vietnamese would be displaced by the encroaching sea. Agricultural and infrastructure damage could total values of nearly £2 billion by the year 2020. This was the consensus alarmist view, echoed by the IPCC report, which predicted climatic catastrophes on a sped-up timetable, although the IPCC scientists spoke merely of 'surprises'.[57]

More sanguine weather commentators, however, said there was nothing new under the Sun. Britain's weather, in particular, was *characterised* by extreme variability: as Bob Hope once complained, 'Where else in the world can you get four climates in one day?'. Some would argue that if Britain's weather history was chequered, America's was naturally apocalyptic. California in 1992 suffered biblical type disasters including fire, flood, tempest, earthquake and riots.

All storms, said Sir John Mason, a former Director-General of the Met Office, were 'within natural variability'.[58] When asked about the meaning of the severe 1990 storm over England he mentioned something about the air over Greenland being colder than usual, but added that there was 'no particular explanation for it. It would be unthinkable if we get the same weather pattern each year'.[59]

Erratic weather, furthermore, could be used to prove either a warming or a cooling. Both Sir John Mason and Michael Hulme are sure there is no real evidence to connect storms with the Greenhouse Effect. Britain in autumn is traditionally windy, even stormy, because of the thermal inertia effect of the oceans which heat up over the warm summer months to release their energy slowly while the land cools more rapidly. Mason and Hulme agreed it is more likely that a warming would reduce storms because the thermal gradient would weaken, and storms would decrease in intensity and frequency because they are created by strong contrasts between cold oceans and warm air.[60]

One of the acknowledged paradoxes of the Greenhouse Effect is that Britain would get noticeably colder because the mild Gulf Stream currents would be deflected further to the south, or would vanish completely. Some, like John Mitchell, head of the Met Office's Hadley Centre for Climate Research, anticipated storms nearer to the British Isles and Europe coming in off the eastern Atlantic, bringing much evaporated warm water with them.[61]

Indeed some thought that colder, stormier weather was quite obviously *evidence* of a cooling. In fact the cooling scenario was adhered to right up to the turn of the 1980s. For example a conference held in 1979 at the CRU in East Anglia had most participants agreeing, in the light of increasingly bizarre weather, that there would be a continuation of the same. Some spoke of a return of characteristics more typical of the Little Ice Age of the 15th century.[62] Others said that the weather of the 1970s and early 1980s was hardly different from the bleak 18th century conditions to which many historical records refer. The common consensus, reflected in the title of a popular science book by Lowell Ponte, was that we were heading, once again, for 'The Cooling'. Indeed, America's popular Science Digest predicted an imminent new Ice Age: 'The present episode of amiable climate is coming to an end ... winter will lengthen year by year, century by century, until it is 365 days long.

Cities will be buried in snow, and an immense sheet of ice could cover North America as far south as Cincinnati . . . '[66].

Already similar articles are beginning to appear again in the popular scientific press.

FOOTNOTE REFERENCES:
Chapter One: Is There a Warming?

1. The Times, 10/2/90
2. Times Higher Education Supplement 30/11/90.
3. Ibid.
4. Ibid.
5. Sunday Times, Hilary Lawson, 12/8/90
6. Ibid.
7. Discover, March 1991.
8. New Scientist, Fred Pearce, 19/12/92.
9. Scientific American, August 1990.
10. Ibid
11. New Scientist, Stafan Rahmstorf, 1/5/93.
12. Scientific American, Robert White, vol 263, July 1990.
13. Sunday Times, Mark Hosenball, 22/4/90.
14. Sunday Times, op cit, 12/8/90.
15. New Scientist, letters, 5/11/94.
16. New Scientist, 15/12/90.
17. Scientific American, op cit.
18. New Scientist, William Nierenberg, 9/3/91.
19. Ibid.
20. New Scientist, Ian Strangeways, 23/11/91.
21. Scientific American, Philip Jones and Tom Wigley, August, 1990.
22. The Times, Nigel Hawkes, 8/11/90.
23. *Living in the Greenhouse*, Michael Allaby, Thorsons, 1990, p113.
24. New Scientist, op cit, 23/11/91.
25. Scientific American, op cit, July 1990.
26. Scientific American, op cit.
27. Ibid.
28. Daily Telegraph, 4/7/92.
29. Ibid.
30. Sunday Times, 31/7/94.
31. *Hothouse Earth*, John Gribbin, Bantam Press, 1990, p7
32. Ibid, p8.

33. Ibid, p11.
34. Ibid.
35. Ibid p14.
36. Geophysical Research Letters, July 1991, James Elsner and Anastasios Tsonis.
37. *Our Drowning World*, Antony Milne, Prism Press, 1989, p79
38. Discover (US), November 1993.
39. Ibid.
40. Scientific American, op cit, August 1990.
41. Sunday Times, op cit, 12/8/90.
42. Times Higher Ed. Suppl, op cit, 30/11/90.
43. New Scientist, 5/2/94
44. Ibid.
45. New Scientist, op cit, 23/11/91.
46. *Future Weather*, John Gribbin, Penguin Books, 1982, p22
47. *Our Drowning World*, op cit, p78
48. Ibid, p79.
49. Ibid, p80.
50. Ibid, p80.
51. Ibid, p81.
52. New Scientist, Paul Simons, 7/11/92.
53. Guardian, 2/10/92.
54. New Scientist, 7/11/92,op cit.
55. New Scientist, 2/2/91.
56. Ibid.
57. Ibid.
58. New Scientist, 16/1/92.
59. New Scientist, 27/8/94.
60. Scientific American, op cit, July 1990.
61. Sunday Times, 4/3/90.
62. Ibid.
63. Ibid.
64. New Scientist, op cit, 7/11/92.
65. *Future Weather*, op cit, p54.
66. Discover (US), November 1993.

Chapter Two:

THE GREAT CARBON RIDDLE

Peter Moore, an ecologist at the Biosphere Sciences Division of King's College, London, wrote in 1991: 'Only 30 years ago *New Scientist* could confidently proclaim that global cycles of elements such as carbon were far too vast to be affected by the activities of humanity. Now the story has changed, and with it the textbooks'.[1] Unfortunately *New Scientist* was probably right, and the new textbooks wrong.

To see why let us remember that carbon chemistry is the basis of all organic life on Earth. This means that our planet is, in effect, a carbon recycling machine, and its main biological features are determined by this. The carbon atom has enormous versatility, and is able to swap up to four of its outer electrons. One atom can combine with four of hydrogen to become a hydrocarbon and by piling on other carbon atoms can create large organic molecules.

It is easy, of course, to see how poisonous carbon compounds can become precisely because of this promiscuous versatility, and how they can alter the molecular balance of atmospheric gases. But one should not exaggerate the consequences of this. Carbon dioxide is a perfectly harmless, odourless gas that exists in minute quantities. At present its concentration in the atmosphere is about 0.03 per cent, commonly expressed as 350 parts per million (ppm).

Not only does this percentage seem paltry to the layman, it is beginning to seem like that to the expert. Indeed it has been alleged that some 100 billion tonnes of carbon have 'gone missing' since the industrial revolution,[2] and that 1.6b tonnes a year of CO^2 simply cannot be accounted for. The search for the missing carbon, rather like the search for the missing mass of the universe, 'is perhaps the most important problem that we have to solve' says Jorge Sarmiento, an ocean modeller at Princeton University. 'It's what

Nature's done to us' adds Richard Houghton, of the Woods Hole Research Centre, in Massachusetts. 'The carbon is there, but nobody can find it'[3]. What Houghton implies is there are enormously complicated feedbacks involved in working out what happens to Nature's total stock of carbon. No clear distinction is made, nor can it be, between the various sizes and significances of the 'sinks', or reservoirs, in the scientific literature.

The difficulty in calculating the nature and speed of the CO^2 recycling process is in the paucity of accurate historical measurements. There is no direct knowledge of the concentrations dating from more than 100 years ago, and even 20th century figures are fraught with hazards. Indeed it could be argued that, judged on the time-scale of the Earth, current levels are actually low. This must be historically true. Prehistoric plants, especially those in the carboniferous period, thrived precisely because CO^2 was many times above the current level, as we shall see. The IPCC, in its 1990 report, was unnecessarily alarmist about the gas, saying its very presence could lead to major changes in the chemistry of the entire carbon cycle.[4] It would, say the authors, bring about significant feedbacks (in this case meaning amplifications of the effects) on CO^2 levels.

The carbon in the atmosphere has a very close symbiotic relationship with geophysical or hydrospheric 'sinks'. Probably about 1,500b tonnes of carbon actually exist in the *ecosphere*, which includes the atmosphere. Carbon is either sunk into the *biomass* (the Earth's vegetation), or into the soil and rocks (as geochemical sinks, which we could call the *geosphere*), *or* into the oceans (the hydrosphere), *or* into the atmosphere alone, or into all four together. CO^2 dissolves in rainwater to form mild carbonic acid, which reacts with limestone and granite to produce calcium and bicarbonate ions. These, plus silica, become embodied in magma, and are eventually released back into the atmosphere during seafloor spreading or when rock undergoes stressful changes.

In other words carbon dioxide is stored everywhere, even in our bodies, and is being continuously added to *and* subtracted from on a seasonal, annual and centuries-long and millenia-long cycle. Any increase in one place within a human lifetime would be too small to measure. The important point to remember is that the carbon cycle is a closed system, as is the hydrosphere. Carbon can increase, of course, as the biomass and the human population increases. But nothing escapes into space or becomes dissolved away, even with molecular dissociation. The particles always get back together at some time or in some way. Earth's volume of water, for example, never increases, because the fumeroles, volcanoes, geysers and springs pump out only what water and vapour they already possess, and it will eventually be transported back into the atmosphere and then recycled down to the surface in rain-storms.

The Origin of Carbon in Pre-History

There is now a growing body of interesting theory placing more emphasis not only on the role the soil plays in the carbon cycle, but also on the internal dynamics of the Earth. Our planet has suffered from the beginning from wild climatic swings, with CO_2 sometimes, but not always, playing a pivotal role. During the first half of Earth's long history temperatures seemed to be much higher than the energy coming from the infant sun could warrant, as it gave out barely three-quarters of the heat and light that it does today. And yet instead of the oceans freezing solid life germinated in Darwin's proverbial warm pond, when the planet was barely a billion years old.

Perhaps, say climatologists, Earth had carbon dioxide levels a thousand times the present level to make up for the solar short-fall, probably from the seething activity of the early volcanoes, which were also responsible for pumping out water vapour in vast quantities that eventually formed the oceans. The faster the conveyor belt of drifting continents moves, the more CO_2 bubbles up from roiling underwater volcanic rifts, all the time with new types of plants varying the rate of CO_2 sinking and emission. Indeed, David Rea of the University of Michigan proved from sedimentary cores of 55 million years ago that underwater hot vents spurted out iron and other minerals at 10 to 100 times today's rate, all coinciding with a global warming.[5]

Writing in 1875 the great founding father of modern geology, Charles Lyell, speculated that if the Earth's surface land had been divided up into smaller regions, Earth's climate would have been more uniform[6]. If there had been mountains higher than the Himalayas 'there would be a greater excess of cold' he wrote in the 12th edition of his *Principles of Geology*.

Lyell's hypothesis now seems highly prescient, since some scientists are suggesting that perhaps the Himalayas are to blame for past ice ages. This novel idea came from William Riddiman and Maureen Raymor while at Columbia University.[7] Tectonic forces succeeded in raising the vast Tibetan plateau some 250 million years ago in the largest continental uplift known to science, when India virtually crashed into Asia. This shifted the Earth's winds and altered rain patterns in Tibet to wash out CO_2 from the atmosphere faster than usual.

The Tibetan plateau covers about 0.4 per cent of Earth's surface, with an average height of about five kilometres above sea level. Maureen Raymor likens the plateau to a giant boulder thrust into the atmosphere, disturbing the circulation patterns of the north. Other computer models suggest that if there was no plateau there would be no Indian monsoon, and it is torrential rain that yields another clue: the higher the mountain range the greater the wash-down effects of chemical erosion, and being situated close to the sea the mountains bring on more precipitation.

But was the weather first the spur, rather than the other way round? Perhaps with intense rain-erosion the sides of the mountains become steeper and the uplifted area floats higher on the mantle. This was suggested by Peter Molnar of the MIT and Philip England of the University of Oxford, although they could not explain the climate change of 50 million years ago. Some argue for positive feedbacks (like more ice-capped mountains in colder climates creating more erosion), and some for negative feedbacks like the idea of *less* weathering when climate gets colder, so restoring the balance.[8]

However, feedbacks and their implications have created enormous intellectual difficulties for scientists trying to determine what happened in early times. Earth, although it may have looked like its sister Venus in the beginning, didn't remain like it. Rocks were subject to organic processes which changed them into soil. Then the first buds of vegetation in turn affected the level of CO_2. To add to the carbon riddle atmospheric temperatures did not always match CO_2 levels in the most obvious ways.

The problem with tectonic uplift events is that they were too often just one-off affairs. Robert Berner of Yale University has done much work on prehistoric atmospheric gases and their various reservoirs. Without a natural equilibrium arising between the sinks and source, he said, life as we know it might not have evolved, or perhaps CO_2 would have disappeared altogether.[9] This equilibrium was largely due to Earth literally becoming earthy, a process which must surely happen on other solid planets if biological and animal life is to evolve elsewhere. Early microbes were to blame. Bare rock simply flaked and granulated and got washed away. But microbes secrete sticky substances that bind particles together and this soil-creation would have been aided, said earth scientist Tyler Volk of New York University, by dissolved CO_2 continuously eating away at the rock particles.[10] Volk and his co-scientist, climatologist David Schwartzman of Howard University in Washington, showed that lichen can break down rock 100 times more quickly than would occur with natural weathering, creating temperature changes of up to 30C.[11] Perhaps, says Robert Berner, it was not the microbes but the early complex plants such as the horsetails, sending down their shoots into the ground, that got the CO_2 cycle going.[12]

Scientists at Columbia University suggest that when the vast Tibetan plain rose up some 250 million years ago it caused an ice age by shifting the Earth's winds and altering its rain patterns. This in turn washed carbon dioxide out of the atmosphere faster than usual.

Source: New Scientist

Crucially involved is the impact of subterranean volcanic eruptions and biochemical processes. Earth rocks that contain carbon consist of two types: kerogen (sedimentary organic matter), and carbonates (skeletal debris of ancient organisms). In the course of chemical weathering kerogen reacts with oxygen to produce CO_2. Carbon weathering is more complicated, involving calcium carbonates being attacked by acids in groundwater. Silicates also produce bicarbonate compounds, and as only half the CO_2 taken up from the atmosphere during silicate weathering is returned as CO_2, there is a net loss of the gas.

This would have so fundamental an effect on Earth that all the CO_2 would eventually be used up were it not for volcanic eruptions. And most degassing of CO_2 arises over tectonic subduction zones, where carbonate sediments are thrust down into the superhot bowels of the Earth. Most soda springs are near subduction zones. The degassing rate is directly proportional to the rate of generation of new sea floor, although it is admitted that because of the rise and fall of sea levels first less and then more land is exposed, affecting the rate of chemical weathering, and this could blur the outgassing picture. Still, there is much speculation about the biological carbon cycle. What fossil clues there are give an indication of prehistoric land forms: for example in the Triassic period, from about 240 to 210 million years ago, there were more deserts.

Much depends, then, on what happens when the continents drift apart, which, of course, they still do. Perhaps the large northern hemispheric continental areas were pushed further north and became more bog-like, and more likely to become occasionally glaciated as they drifted further towards the poles. Again the feedback: after each glaciation the landscape looked more boglike: 'glaciations have gradually remodelled large parts of the northern hemisphere into types which are more suitable for wetland formation', said one scientist.[13]

Whatever was going on, CO_2 levels of olden times were truly extraordinary. Robert Berner, who is also an expert on ice geochemical carbon cycles, ran computer models to explore how CO_2 might have changed over the past 600 million years. Some 500 million years ago, he reckoned, the CO_2 level was 18 times higher than at present. However there does seem to be much controversy about the later Cretaceous period, between 144 million and 65 million years ago, both in regard to the level of CO_2 at the time and concerning atmospheric temperatures. In fact a graph included in a *Scientific American* article co-authored by Berner shows an astonishing level of carbon dioxide some 100 million years ago of more than 30 times the present level.[14] And yet elsewhere Berner has said that the Cretaceous atmosphere contained about four times the present level of CO_2.[15]

What is important in all this for the present warming debate is that temperatures seem to be a disconnected variable, perhaps because prehistoric episodes of warming and cooling had something to do with other geophysical events, or the sun itself. Late in 1994 Paul Valdes and Brian Sellwood repeated in a letter to *New Scientist* that any warming in the Cretaceous was 'almost certainly' not only the result of CO^2, and that 'CO^2 is not the only factor that affected climates in the distant geological past'.[16]

Indeed, the Cretaceous itself seems to be a hotbed of mysteries, and resolving the carbon dioxide issue may finally provide the clue as to what happened to the dinosaurs, the nature of evolution itself and the final destiny of our planet. Rowan Sage, a physiological ecologist at the University of Toronto, revealed in 1994 that 135 million years ago the air contained only five to ten times as much CO^2 as it does today. This of course conflicts with the opinion of other scientists, as we have seen. But Sage nevertheless made the interesting observation that the reason the giant sauropod dinosaurs were so large was because of the low protein yield of Cretaceous vegetation. Because of high levels of CO^2 plants like cycads had only half the protein levels of modern-day plants. To get enough protein a brontosaur had to eat heroically. This in turn was due to the shortage of nitrogen in Mesozoic plant life owing to the virtual absence of the plant enzyme rubisco, a nitrogen-rich molecule, in such high CO^2 levels.[17]

At the height of the last ice age, roughly 18,000 years ago, CO^2 was hovering at record lows of around 200 ppm, compared with 355 ppm today. By that time plants had stocked up with the high rubisco levels we see today. When the ice age came to an end about 12,000 years ago CO^2 rose to about 270 ppm, says Sage.

The 'Missing' CO^2

The general consensus is that five billion tonnes of carbon exists today as atmospheric CO^2. There should still be more. The sinks, whichever they are, seem to be more efficient at absorbing the gas than current theories explain. Oceanographers such as Wallace Broecker of the Lamont-Doherty Institute, say most of the carbon finishes up in the sea, and I will have more to say about the oceans in Chapter Four. Geochemists such as George Woodwell of the Woods Hole Institute say it is all being buried in the biomass. Environmentalists say there is 'no guarantee' the sinks will continue to absorb ever-increasing amounts of man-made CO_2. The division of opinion seems to be between optimists who suggest they may well be spurred to ever more absorptive efficiency, while the pessimists who say that the sinks by now must be full up, and a 'climatic flip' or 'runaway greenhouse effect' is just around the corner.

However, the missing CO_2 must cast serious doubt on the 'diminishing biomass' theory, because as vegetation shrinks less CO_2 is sunk and oxygen levels fall. Again, either vegetation is not declining, or the invisible sink is working overtime. So is the world's stock of vegetation decreasing, increasing or remaining the same? The remorselessly depressing doomsday image is one of Man slowly but surely reducing the size of the forests and the world's stock of vegetation, releasing back into the atmosphere billions of tons of carbon that have taken literally millions of years to accumulate.

The burning of biomass is thought to be a serious matter, because there is probably as much carbon in the top yard of soil as in the biomass and atmosphere combined.[18] Hence the clearing of the topsoil, as a concomitant of deforestation, seems to do most of the damage because living organisms, including plants, decompose at or just below the surface. All this means that the soil 'reservoir' of oxidized carbon is suddenly released into the atmosphere.

The evidence of loss of CO_2 as a result of Man burning fossil fuels, rather than other biological processes having sunk CO_2, in the last 100 years, comes from ice cores which show radiocarbon decay due to the great age of coal deposits. These deposits contain lower carbon-13 isotopes than the present carbon-14 in the atmosphere.[19] However, some scientists deny the validity of checking air bubbles trapped in polar ice cores. They say it is an inaccurate method of working out the level of ancient CO_2, since the ice cores can be contaminated by recent atmospheric air during the process of recovery. A group of Norwegian and Japanese polar research scientists say: 'It is astonishing that these [ice cores] studies have been so credulously accepted ... The ice in polar sheets is not a single, solid-state phase with bubbles in which the air is preserved indefinitely'.[20]

However the most important handicap for students of the greenhouse effect is the lack of conformity in the use of notation. Seldom do academic papers use the same data-base, the same quantifying techniques. Tonnages of carbon burnt, stored, emitted or recycled are never the same for any two scientists. We have seen already the wide diversity of opinion about the amount of CO_2 that existed at the time of the Cretaceous. Annual CO_2 emissions due to forest destruction is estimated at '2.2 to 10.2 billion tonnes — 10 to 50 per cent of fossil fuel emissions', according to environmentalists Stewart Boyle and John Ardill.[21] But is destruction the same as burning, and how much should be allocated to each?

A good example of the quantification difficulties can be shown by the reception that the book *Global Biomass Burning: Atmospheric, Climatic and Biospheric Implications* received. This book was published by the MIT Press in 1992, and gathered together the research findings and opinions of 158 scientists from around the world; it could be reckoned to be a definitive sequel to the IPCC report. The papers were presented to the American Geophysical Union, and put into perspective many of the fears and assumptions of the Intergovernmental Panel.

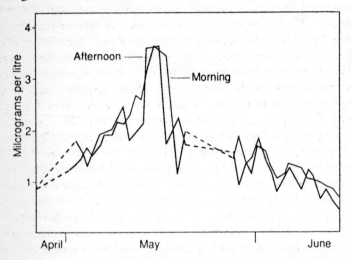

Knowing how much carbon in the ocean is recycled and consumed by micro-organisms on a seasonal basis is crucial to our understanding of natural climate-controlling mechanisms. The amount of phytoplankton (measured in terms of the abundance of chlorophyll) differs over time, but peaks in the spring bloom. It can take a thousand years for recycled carbon to reach the ocean surface.

Source: New Scientist

This book was not easily obtainable, and most interested parties would have had to rely upon an extensive review of its contents published in *New Scientist* in September 1992 by Peter de Groot, a scientist at the Science Council of the Commonwealth Secretariat. Unfortunately de Groot merely confuses in his attempt to convey the uneven message of the book's contributors. Biomass tonnage and 'tonnes of carbon' seem to be the same; i.e. 3.7m tonnes of savanna biomass burned, and 3.7m tonnes of 'carbon' goes into the atmosphere. But this cannot be right, because when one tonne of carbon is burnt, about four tonnes of CO_2 are produced.[22] Tropical forests contribute some 25 per cent to the global greenhouse, de Groot learns. And yet 'best estimates suggest that fires consume nearly 3.7m tonnes of savanna, and almost 1.6m tonnes of forests ... on a dry weight basis'. But 1.6m tonnes is much more than 25 per cent of 3.7m tonnes. Earlier de Groot wrote that savanna burning releases 'three times more carbon than is released through forest burning'.[23] He continues by saying that the 'general biomass' burned produces 3.5m tonnes of CO_2, so already we have lost 200,000 tonnes of the gas in the space of one inch of type. Furthermore Philip Fernside of the National Institute for Research in the Amazon calculates that Brazil emits nearly three times as much CO_2 from rainforest burning 'than from the use of fossil fuels', which is a 325 per cent increase on de Groot's figures. Such quantities 'may be as large as the anthropogenic sources of sulphate aerosols', says Jenner.[23] Even so, we are led to understand that the combusted biomas produces 'up to 40 per cent of the world's annual CO_2'.[24]

According to Joyce Penner, a climate researcher at the US Government's Lawrence Livermore National laboratory, deliberate burning of grasses and forests spews around 50m tonnes of organic aerosol into the air each year, compared with 4m tonnes from natural fires. But this doesn't tally with the general consensus that humans emit 5 *billion* tonnes of carbon dioxide annually into the air.[26] And this in turn is not to be confused with the 4.6 billion tonnes emitted solely from forest destruction 'assumed by most climate modellers'.[27]

Other academic sources say that the clearing of Earth's vegetation, especially by slash-and-burn cultivation techniques in the savanna and grasslands (for, say, cooking, heating and to facilitate hunting), is not apparently happening on any alarming scale. Biologists may have over-stated the severity of such techniques, and may have wrongly believed that Third World peasants have not allowed the biomass to recover. We can virtually rule out deforestation on a grand scale leading to desertification. Ulf

Hellden, head of the Remote Sensing Dept at the University of Lund, Sweden, says even international papers challenge the concept of Sahelian zone desertification, largely due to the lack of data. He quotes official statements by prominent figures such as the president of the World Bank, George Bush and others, that have varied the rate of desertification from five to 90 kilometres a year. He says such statements are rarely accompanied by descriptions of the survey techniques and methods used. His research in the Sudan since the 1960s, using field surveys and satellite photos, show no significant desert encroachment between 1962 and 1984.[28]

In fact, Africa may contain twice as much wood as previously believed, according to a British study for the World Bank, published in June 1994, and a similar calculation was made earlier by the UN's FAO. 'The crisis of fuel wood on the African continent may have been seriously exaggerated', reported Phil O'Keefe of the University of Northumbria at Newcastle, who co-ordinated the World Bank report.[29] Other findings suggest that the temperate forests are not so much 'vanishing' as adversely affecting wildlife and biodiversity. Indigenous tree species are not being demolished for good, but are being replaced with newer types of plantation to supply the pulp industry. A World Wildlife Fund report in October 1992 referred specifically to Chile, where nearly 20 per cent of the country's forests are under plantation with a newly introduced Monterey pine, while much of the native forests were exported.

This is not to deny that the changing nature of Europe's deciduous and native coniferous forests is not a serious aesthetic and environmental problem. Only eight per cent of Britain is forested, and much of what was once Caledonian pine forest in Scotland is now planted with Sitka spruce or lodgepole pine which 'have not adapted to the ecology of Scotland'.[30] Much of the Earth's timber supply, of course, is 'harvested', and in fact more saplings are planted than mature trees are cut down. There are now desultory attempts in Asia and Latin America to get loggers to replant trees in the areas they have cleared. In Europe the pulpwood industry, especially SCA in Sweden and Shotton Paper in Britain, uses conifer trees, especially Sitka spruce. SCA alone plants 60 million seedlings a year, and the Forestry Commission says that trees are now being planted in Britain on nearly 100,000 acres a year.[31] In 1992 Finnish forest researchers estimated that European forests increased in terms of acreage covered annually by one per cent during the past quarter of a century.[32]

Biomass burning, it seems does *not* do three things — reduce, long-term, the amount of vegetation; have any bearing on the missing carbon; or make the planet warmer. Indeed, some point to the natural role that fires and hot volcanism have played in evolution in creating and controlling atmospheric gases. By the later Tertiary period, a combination of fire, herbivores and climate had largely shaped the savannah landscape from where our ancestors are said to have evolved.

Other authors cited in the *Global Biomass* review imply that CO_2 from biomass burning is hardly a problem, since other gases in equal or greater quantities are emitted. Meinrat Andreas of the Max Planck Institute says that the burning produces 32 per cent of carbon monoxide, 21 per cent nitrogen oxide, and 25 per cent hydrogen, 'and is responsible for the production of 38 per cent of ozone in the troposphere'. Hardly surprisingly de Groot speaks of the 'awesome complexity' behind the data, and he admits to the uncertainty of our knowledge.[33]

Furthermore, as carbon dioxide stimulates plant growth by increasing both the rate at which plants fix carbon and the efficiency with which they use water, Sherwood Idso, of the US Water Conservation Laboratories, believes that burning of fossil fuels on a grand scale would actually be beneficial for existing biota.[34] For example in the late 1980s Idso did experiments on sour orange trees with double the normal CO_2 dose. These trees are now twice the height of a control group of trees, and have 180 per cent more biomass. Fifty years ago similar findings were made by nursery owners who used CO_2-enriched air to enhance crop production. Certainly tropical deforestation is not releasing as much CO_2 into the air as was thought, largely because of the neglected role of CO_2 absorbed by vegetation regrowing on cleared land.

Other modifying changes in the ecosphere concern peat bogs, and often such bogs coexist with forested areas. Lee Klinger of the US government's National Centre for Atmospheric Research (NCAR) in Boulder, Colorado, claims that peat bogs are a giant natural store of carbon, holding between 500 and 1,000 billion tonnes of the element — more than all the world's trees and similar, he says, to the 760 billion tonnes held in the atmosphere.[35]

Lars Franzen of Gothenburg University thinks peat bogs have accumulated more than 200b tonnes of carbon in the 10,000 years since the end of the ice age. Antarctic ice cores show past temperatures, plus carbon dioxide and methane. So the creation and decomposition of the bogs must explain changes in temperature — they are massive CO^2 sinks.[36] Peat is a form of moist, partly decomposed plant material, on its way to becoming coal deposits. Peatlands cover about 5m square kilometres of the Earth, ranging from the tropics up to the frozen tundra of Siberia, going down to a depth of 60 feet in places, holding up to 100 times as much carbon per hectare than tropical rainforests. Cooling encourages the growth of peat bogs. They extract carbon from the atmosphere and store it, so perpetuating the cooling (since warming tends to release CO^2 into the air largely through convection). Klinger claims that peatlands could have been an important biological mechanism that could have plunged Earth into and out of ice ages,[37] by changing the amount of CO^2 in the air.[38] Richard Clymo of London University also points out that there are remains of birch woods dating back some 5,000 years which have been overwhelmed by peat.[39]

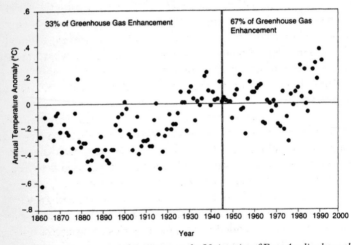

Data from the climate research institute at the University of East Anglia shows that almost all of the warming this century came before most of the greenhouse gases were emitted.

Source: Patrick J. Michaels, Cato Institute, Washington, DC

Fossilized leaves of an Arctic willow species from peat deposits in Norway and Scotland going back 100,000 years also point to a longer time span over which leaves adapt their pore density. Plants tend, in other words, to reduce water loss through their stomata rather than take up more food from the CO_2.[40] It seems that leaves grown at low CO_2 levels form more stomata than when exposed to higher concentrations. The British Museum, says one scientist, has plant specimens going back to the 19th century with more stomata (gas exchanging spores) than those today. This suggests plants have become more water-efficient today rather than that CO_2 levels are up.[41]

The Biomass and Current CO_2 levels

Much of the foregoing hints that the biomass is not decreasing to the extent once thought, and I have concluded that CO_2 is not accumulating as a result. But do we know for sure? Scientists are still bidding at the auction. Some like Jorge Sarmiento of Princeton University, believe the forests play a large role in the sink. Columbia University climatologist Aiguo Dai ran a computer simulation based on Met records going back 50 years. He found that forests could be soaking up about half the 1.6b tonnes a year of missing CO_2.[42] This compares starkly with the findings of Finnish researchers who concluded that the European forests had been absorbing as much as 120m tonnes of carbon a year, and probably reflected what was going on elsewhere in the world's biomass.[43]

Tom Wigley, formerly of the CRU and now at the NCAR, says that land vegetation plays a largely neutral role.[44] Forests have grown at variable rates in the past 100 years, absorbing sometimes more, sometimes less, of the gas. Canadian and Siberian forests, says Richard Houghton of the Woods Hole Research Centre in Massachussetts, are mature and slow-growing, unable to capture much CO_2.[45] Perhaps they soaked up more in the past. In the meantime, he says, European forests are still young and growing.

Even so, it was revealed in 1993 that data buried for years in forestry ministries tended to support theories that suggest northern (coniferous) forests are no longer soaking up CO_2 to the extent they did.[46] This data included girth size, spacing etc, and hence the volume of wood. The forests seemed to grow rapidly between 1920 and 1960 which some have attributed to carbon dioxide enrichment.

The evidence concerning modern boreal forests is no clearer. Before 1890 boreal forests were a source of carbon dioxide because there was widespread felling, which caused debris from the forest floor to rot more quickly. Then in the late 1970s forests began to loose wood again and became a source of CO_2, taking over partially from the CO_2 enrichment due to man-made activities. The net balance is probably zero.[47]

Allan Auclair is a plant ecologist working with the environmental consultancy Science and Policy Associates in Washington. He has done estimates of tree growth rates in uncut forests, derived from studies of tree ring width over the past 100 years in northern forests. He concludes that the forests have changed from being a carbon source to a carbon sink and back again.[48]

Perhaps there is a case for saying that tree biomass has been expanding, and this has been due not only to an increase in atmospheric CO_2 but also because of the extra availability of other soil nutrients such as nitrogen oxides, phospherous and water, which in turn are the result of the warming. In other words the biomass as a whole has to be taken into account,[49] with cleared forests growing back in some areas matched by other arboreal and biomass changes elsewhere.

Some scientists think that the role of soil microbes has been under-estimated. Some of the released CO_2 is taken up by micro-organisms and pushed back into the Earth. In crude terms one could argue that the self-regulating Gaia principle is at work. The key question, however, is whether the northern forests have somehow got ahead of the microbes, and are succeeding in converting CO_2 into biomass faster than microbes can send it back.[50] Columbia University climatologist Aiguo Dai thinks that plants have gained over microbes because of a change in the weather.[51]

Indeed it seems that soils from the tropics have microbial biomasses less sensitive to temperature increases than in northern European soils.[52] Overall global changes in carbon in soil and microorganisms would be difficult to confirm empirically, as they evolve over tens of years. An attempt, however, to do this was made by scientists at Reading University. They developed, in conjunction with co-workers at Rothamsted, a Carbon Model to take into account the effects of agricultural management to predict how much CO_2 will be liberated from the world stock of soil and organic matter for any given temperature rise. The model predicts that over the next 60 years about 61b tonnes of extra CO_2 will be released from the soil, and nearly 20 per cent will be released from soil organics as a result of the warming.[53]

If soil micro-organisms can reinforce a warming, it supports the theory that growth factors are less determined by CO_2 enrichment than latitude temperatures and the availability of nitrogen in the soil.[54] Microbes feed on nitrogen and phospherous, and discharge considerable amounts of CO_2 and nitrous oxide and methane. But the warmer soils could increase the 'hunger' of the microbes, spurring a further warming.

In conclusion, until scientists have fully worked out all these complicated feedbacks, we may assume that: 1) there is no destruction of the biomass on any sizeable scale; and 2) the level of CO_2 is subject to self-regulating variables that are independent of the former; and 3) any additional anthropogenic emission of CO_2 is not accumulating in the atmosphere, but is being rapidly 'sunk' into Earth's various natural reservoirs.

Perhaps the conundrum concerning the present-day environment could be resolved if one were able to look at the Earth from space with a device that could detect the natural 'breathing' patterns — to see which parts of the ecosystem take in CO_2 and which parts emit it. This has actually been done, and many scientists have been surprised to observe that the net absorbers are *greater* than the net emitters. Researchers at Nasa's Ames Research Centre in California in 1993 used a model known as CASA, driven by readings of infrared light taken by the NOAA-9 satellite which, when combined with temperature and sunlight readings, give an estimate of how much photosynthesis is taking place. The model also takes into account the type of vegetation and soil texture typical of regions of the world.[55] The satellite showed clearly that the net absorbers of carbon dioxide spread across most of the equatorial regions of Africa, Siberia, eastern Europe, Amazonia and northern Canada and the north-east of the United States, and parts of the far east.

Industrialisation and CO_2

I have concentrated overly on the natural carbon cycle because this is where the main sources and 'sinks' of carbon dioxide reside (the oceans are also a major sink, and I will refer to them in Ch 4). Industrial emissions are only one out of five other natural biospheric sources of the gas, so in theory we should be more fearful about what is happening to the natural environment than the man-made environment.

However it is true that one of the major worries of many climatologists is the effect of industrial emissions of CO_2. There is no doubt that there has been an increase in warming potential as a result of industrial processes in the northern hemisphere, emanating mainly from the burning of fossil fuels in factories, power stations (where the chief energy source is coal), industrial plants and car exhausts.

A graph from the famous Mauna Loa Observatory, taking samples from the air from a site perched on top of a volcano, show definite rising levels since the beginning of the industrial revolution (from 277 to 351 ppm). Scientists are concerned not only about this clear exponential trend, but about the fact the gas molecules can remain in the atmosphere for anywhere between 50 and 200 years. And the emerging Far Eastern and former Soviet countries are increasing their percentage of the total from an already high base, especially the latter where a higher proportion of their economic output has traditionally been from the 'smokestack' kind of industries.

However two important reservations must be borne in mind. The first is that CO_2 emissions show a *declining* trend in relation to gross national product (GNP). For example, the 1988 global average emission is about 0.35 metric tons of CO_2 for every \$1,000 of GNP.[1] But the United States, with its very high rates of CO_2 emission for every person, is actually *below* this world average. The implication is that as the other emerging economies slowly reach American levels of energy-efficiency their carbon dioxide emissions will drop accordingly.

The other point is the conundrum arising from the absence of any catastrophic warming that should surely have arisen by now given the exponential rate of rise in warming gases. There has, after all, according to the IPCC report, been a 50 percent increase in CO_2 since the industrial revolution, and a further 40 percent increase over the past 100 years. Surely the global temperature should have gone up by much more than the one or a half degree often reported? As it hasn't, why worry about its exponential nature? Furthermore computer models showing hypothetical further doublings of CO_2 to 600 ppm from six different research institutes range from less than 2C to more than 5C.[2] It is hence clear that the models do not come up with similar data, and there is no exact mathematical relationship between atmospheric temperatures and CO_2 concentrations, something we have already learned from our prehistoric examples.

FOOTNOTE REFERENCES:
Chapter Two: The Great Carbon Riddle
1. New Scientist, 12/10/91.
2. Discover (US) December 1993.
3. Ibid.

4. New Scientist 26/5/90
5. Discover (US), November 1992.
6. New Scientist, 3/7/93.
7. Newsweek (US) 23/11/92.
8. New Scientist 3/7/93, ibid.
9. Ibid.
10. Discover (US),November 1992.
11. New Scientist, 2/9/89.
12. Discover, ibid, November 1992.
13. Ambio, vol 23, p300, December 1994.
14. Scientific American, March 1989.
15. New Scientist, 21/12/90.
16. New Scientist, 5/11/94.
17. Discover (US), October 1994.
18. Guardian, Phil Brookes and David Jenkinson, 3/4/92.
19. *Hothouse Earth*, John Gribbin, Bantam Press, 1990, p105.
20. The Science of the Total Environment, vol 114, August 1992.
21. *The Greenhouse Effect*, Stewart Boyle and John Ardill, New English Library, 1989, p34.
22. Hothouse Earth, ibid, p98.
23. New Scientist, 12/9/92.
24. Ibid.
25. Ibid.
26. *Turning Up the Heat*, Fred Pearce, The Bodley Head, 1989; see also New Scientist, 4/5/91.
27. *The Greenhouse Effect*, ibid, p34.
28. Geographical Magazine, May 1992.
29. New Scientist, 11/6/94.
30. New Scientist, Adrian Barnett, 7/11/92.
31. *Our Drowning World*, Antony Milne, Prism Press, 1989, p57.
32. Discover (US), December 1993.
33. New Scientist, ibid, 12/9/92.
34. Sunday Times, Hilary Lawson, 12/8/90.
35. New Scientist, 3/4/94.
36. Ambio, vol 23, p300, Ibid, December 1994.
37. New Scientist, ibid, 3/4/94.
38. New Scientist, 3/12/94.
39. New Scientist, ibid, 3/4/94.
40. Times Higher Education Supplement, 7/5/93.
41. New Scientist, Ian Strangeways, 23/11/91.

42. Discover (US), December 1993.

43. Ibid.

44. New Scientist, 11/9/93.

45. The Times 27/3/95.

50. Sunday Telegraph, Matt Ridley citing *Nature*, 30/1/94.

51. The Times 27/3/95.

52. Time magazine, 20/3/95.

53. Sunday Times, 26/3/95.

54. Discover, ibid, December 1993.

55. Ibid.

56. Guardian, 3/4/92.

57. Ibid.

58. *Turning Up the Heat*, Fred Pearce, op cit., p118.

59. New Scientist, 8/1/94.

60. Robert C Balling, Jr, *The Heated Debate*, Pacific Research Inst.,
     1992, p25.

61. Ibid. p39.

Chapter Three:

THE COOL SKIES

The ultimate source of all heat, of all atmospheric moisture and indeed of most atmospheric gases, is the sun. Its radiation, arriving on a variety of spectra, is channelled into the atmosphere which acts as a complicated maze of filters by which its heat and energy is distributed unevenly across the surface of the Earth. Pockets of air rise and fall, expand and contract, drift across vast latitudes, fill with moisture and at times become opaque with pollution. As the Earth's surface is warmed by the sun and by man himself, air rises and enters less dense air, and this allows it to expand further. It begins to lose energy, and cools at the rate of about 10C for every km gained in altitude, and this explains why high mountains in hot countries are capped with snow.

The expanding, cooling air reaches an upper limit of the troposphere, that lowest part of the atmosphere dominated by heating from the ground surface. It extends to about 11 miles above the equator and, say, four miles over the poles. The polar temperature is extremely cold — about minus 50C. The layer above this, the stratosphere, reaches up to about 30 miles, but air temperature no longer falls as air ascends, because it is warmed directly by the sun. In fact at the upper boundary, the stratopause, the temperature gains to reach OC (32F). Above this is the mesosphere, where temperatures start falling again to drop, at 50 miles up, to minus 90C.

Beyond this lies the thermosphere, where air temperature increases again, being bombarded with cosmic and solar radiation that starts to strip electrons from atoms, giving them a virtual electric charge. Higher still, about 186 miles up, the atmosphere becomes the ionosphere, so-called because what few gaseous particles left become ionized, that is virtually *all* of them lose their electrons. There are more layers above that, but to all intents and purposes the atmosphere has come to an end.

The troposphere is where all the action is, with warm air rising and colder air descending, creating weather turbulence. Greenhouse gases trap radiation as well as radiate it in almost equal proportions, but it is a complex business. Only gases with chemical bonds can be called warming gases, which automatically excludes single atoms of oxygen, argon, helium, krypton and xenon. Nor can gases in which there is a single chemical bond absorb infrared (IR) radiation (like N, O, H and so on). And we are left with the *molecules* such as water (H_2O), carbon dioxide (CO_2), methane (Ch_4), nitrous oxide (N_2O), ozone (O_3) and a few others.[1]

The quick turn-around that carbon-based gases can make for no apparent reason even surprises scientists. Paul Novelli of the NCAR wrote in the April 1994 issue of *Science* that carbon monoxide levels seem to have come down smartly in the previous three years, along with methane and nitrous oxide, with even CO_2 being stable.[2] Possibly, he surmised, the eruption of Mount Pinatubo in the Philippines in mid-1991 or recent El Nino events were the cause, the latter perhaps changing the direction of winds along with the Pacific currents. Novelli was encouraged to note that the main cleanser of atmospheric pollutants was the hydroxyl radical (which tends to oxidise them). Carbon monoxide has been the largest user of hydroxyls, so its decline will mean that other pollutants will get more of a chance to get oxidised. Other scientists suggest CO_2's decline may be due to tighter pollution control on vehicle exhausts in America, or possibly a reduction in biomass burning in the tropics, or maybe other ill-defined chemical and oxidising processes in the atmosphere.

In the meantime Sir John Houghton of the IPCC was stung to join in a recent heated discussion in the *New Scientist* about the radiative properties of the carbon atom. He said the scientific theory behind it had been understood for many decades. A CO_2 molecule distributes radiative and kinetic energies to create a situation known as 'local thermodynamic equilibrium'. The molecules absorb and emit photons entering and leaving the atmosphere, and this activity can be calculated to work out the net energy input and output into the climate system[3]. 'Methods', he wrote, 'for carrying out these calculations with good accuracy have been in place for over 30 years. No reputable scientist working in the field doubts that the methods are correct'.

Unfortunately they do. In the first chapter I mentioned that the Earth cools by surface evaporation rather than radiation. This is because one of the most important features of ascent and descent physics is *water vapour*. In effect, all gases are literally overwhelmed by Earth's moisture, and virtually all their physical properties are grossly distorted, and even destroyed, in the process.

Air contains 350 parts per million (ppm) of CO_2 compared with the enormous amounts of water vapour which range between one and four per cent (10,000 to 40,000 ppm), depending on temperature and humidity. According to Keith Beyer of the Centre for Global Change Science at MIT, water accounts for 99 per cent of the greenhouse effect that warms this planet by as much as 30C.[4] Water vapour is not important solely because of its much wider radiation absorption characteristics,[5] but because of the way it reacts to surface temperatures and atmospheric pressures. When warm, dry air moves over the surface of water, moisture evaporates into it. As the temperature falls the vapour condenses out of it, creating clouds, and ultimately rain.

But nothing else gets 'condensed out'. Virtually everything is trapped within the tropopause, and is unable to rise further because above this it is warmer still. This includes dust, smoke and polluting grains. These particles themselves play a part in the cloud-reflectivity scenario. Particles can form the nucleus of a new water droplet, and the more there are the whiter the clouds. A molecule of water vapour remains airborne for just over a week, with particulate matter for about two weeks. However, oxygen and hydroxyl molecules (the O-H of H_2O) will react to almost anything, so have the effect of cleansing the air by enabling them to be washed down to Earth by rain.

However, some scientists suggest that the modest warming caused by CO_2 in turn releases more vapour — the typical positive feedback.[6] Yet such a feedback, in truth, would be ineffectual compared to the effect of powerful solar heat and high pressure weather systems. Even doubling the amount of CO_2 would only increase the amount of heat trapped by an average of four watts per square metre, whereas the entire globe radiates 390 watts.[7]

Robert Charlson, an atmospheric chemist at the University of Washington in Seattle, and his colleagues, say this might explain why computer models suggest a greater warming than has actually been experienced, because of the theoretical characteristics of CO_2

modelling.[8] True, CO_2 *is* an efficient heat absorber, but only at two narrow wavelengths. The carbon-oxygen bonds of CO_2 interact only with photons of IR light of exactly the right vibrationary frequency, and it is the vibration that bounces back the heat into space. The IR spectrum spans the EM wavelenth from around one to 40 micrometres (millionths of a metre). CO_2 absorbs wavelengths at two narrow micrometre zones. Furthermore the way they do this depends on their position in the atmosphere. It is a highly complex process, because the effect of an increase in any one is dependent upon all the others.

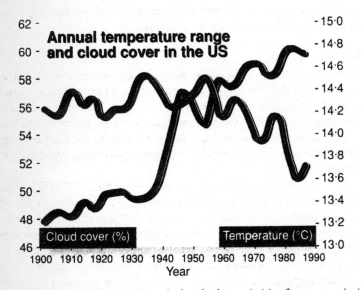

Gaia at work: Earlier disputes about whether cloud cover (arising from a warming) continues to warm the Earth or instead cools it have now largely been resolved.

Added to this complexity is the fact that the absorption bands are stretched almost to their limits, so doubling the CO_2 is not going to increase the absorption of IR radiation to anything like the extent imagined. Hence any continued 'global warming' is more likely to be caused by gases with absorption capacities functioning at other wavelengths. Jack Barrett of Imperial College, London, wrote in October 1994 that CO_2 only releases its heat energy after colliding with other molecules. The half-life of a carbon dioxide molecule excited by IR radiation is 10 microseconds, during which time it will collide with 10,000 other atmospheric molecules. As a result all radiation is totally absorbed, except that passing through a narrow 'window' between wavelengths of 7.5 micrometres and 14 micrometres through which excess heat escapes into space.[9]

Often the term 'greenhouse effect' is interchanged with 'carbon dioxide', thus subtly substituting a physical argument with a chemical one, so the warming feedbacks themselves become operative rather than the dynamics of the airborne gases. But the physical side of the equation remains important. James Walker of the University of Michigan considered that this planetary *weathering* thermostat was the key. A massive feedback is at work with first a warming, then torrential rainfall, then chemical weathering, then more CO^2 distillation, and then a cooling.[10]

Pollution Cools

One of the most remarkable scientific U-turns in recent years is seen in the way that the 'pollutant-as-cooler' argument has gained ground. Atmospheric chemists are now saying that those pollutants that manage to drift ever higher into the upper stratosphere can remain there for up to five years, being warmed by the sun but at the same time prevent the sun's rays from penetrating fully to the ground.

In fact, meteorologists from the Max Planck Institute for Meteorology in Hamburg put the cooling over the past century at more than 0.5C over much of Europe.[11] Dust brought about by over-intensive farming practices is also a factor, and probably contributes about 2b tonnes a year, according to Robert Duce of Texas A & M University, with an unknown amount arising from non-farming human activity.[12] The winds each year also carry off more than 200m tonnes of Saharan dust and waft it across the Atlantic on huge thermal currents[13]. Calculations recently published by Thomas Karl of the US Government's National Climatic Data Centre in North Carolina, suggests that dust should block out heat so effectively that a warming might well be put into reverse in the world's most populous countries. At the same time other areas have warmed in postwar years, while yet others with the highest sulphur dioxide emissions have cooled.[14]

Furthermore, over much of the northern hemisphere, and centres of population such as Australia in the south, raincloud cover has increased substantially. In the US it increased by nearly 50 per cent in pre-war years, about 60 per cent since the mid 1950s. Similar data have been published in *Geophysical Research Letters* for cloud cover in China and regions of Europe.[15] Jeffrey Kiehl of NCAR at Boulder says that summer aerosol cooling would 'completely offset' greenhouse forcing across Europe, especially the eastern parts.[16]

The *sulphur* atom seems to be the main culprit. Research by Mai Pham and Guy Brasseur, also from NCAR, conclude that a global sulphate cooling is in the immediate offing.[17] An international study of daily temperatures from China and in Russia and America shows that night/day temperatures are evening out because of sulphate cooling.[18] This view was supported by Robert Charlson who has done much research on aerosol cooling. He says aerosols degrade scenic views because of light bouncing off them at an angle.[19] Skies in the eastern US are often whitish due to photon scatter. This, he says, is due to sulphur particles which now exceed 2,000 ppb,[20] from smokestack industries, power stations and from car exhausts.[21]

The discovery of the iniquitous effects of sulphur have arisen partly from concern about acid rain in the 1980s. Out of this problem came better techniques for measuring sulphate emissions and better computer models (especially Swedish) of wind patterns and atmospheric chemical mixing (see Chapter 6). Charlson invented an optical scattering device called a nephelometer, based on the electric eye principle. Fed into models, it found that a watt of solar energy per square metre was prevented from reaching the surface — enough to cool the Earth substantially. Hitherto warming models did not take into account the haze effect. But because of their limited range (unlike CO^2) most sulphate emissions float in the northern hemisphere. This tends to increase thermal gradient differences between the north and south.[22]

Volcanoes also emit sulphur into the stratosphere, but anthropogenic sulphates amounting to 90m tonnes a year[23] falling back to Earth soon affect the lower atmosphere below 36,000 feet in middle latitudes. They seem seriously to affect higher latitudes as well, and it is said that pollution over the Arctic is acting like a giant sunshade, cancelling out the expected warming in the region, according to data from 22 monitoring sites. Over the past 40 years Arctic sunlight has decreased by 15 per cent.[24]

The cooling could be yet further exacerbated when particulate matter becomes good condensation nuclei and helps create water droplets that make up clouds.[25] The more aerosols the smaller and brighter the droplets, and the more reflective they are the more they perpetuate a cooling. In fact warming gases have to fight with the increasing levels of pollutants that environmentalists seem to have downplayed in terms of the climatic debate. Says Charlson: 'Since the days of the Nixon administration there have been people saying that if we learn how to pollute just right, everything will be fine'.[26] Whereas the evidence for a warming is an abstract and uncertain one, the evidence for a cooling is much more tangible. Most people in industrialized societies can see with their own eyes the pollution haze that seems to hang in the near horizon virtually all the year round.

However, 'we have partly counterbalanced one form of pollution with another, but that doesn't bring us out of the woods', says Bruce Briegleb of NCAR in the most recent (December 1994) assessment of the problem.[27] It not only makes climate modelling more difficult, but the combination of the two could yet have unforeseen climatic consequences. The two types work in fundamentally different ways, so the cancellation won't be neat. Humans add something like 75m tons of sulphur from sulphur dioxide into the atmosphere each year. Scientists have ignored the problem for so long because they don't know how to measure the effects.

The problem is similar to that of the carbon cycle — working out where it is and where it goes. Walter Ellsasser of the Lawrence Livermore Laboratories, in California, concluded that a doubling of CO_2 might contribute to a cooling, if it does anything.[28] Nor, as some think, is the sulphate cooling especially localized. A recent *Nature* article about computer modelling of sulphates suggested that the cooling might be more widespread than thought.[29] Karl Taylor and Joyce Penner of Lawrence Livermore modelled CO_2 and sulphate aerosols together. The CO_2 warmed, but sulphates cooled in a patchy mosaic, mostly affecting the northern hemisphere. Temperatures, according to the model, dropped by 2.2F. Even in the south the temperature went down to a respectable 1.4F (0.8C). 'The climate doesn't change just where the aerosols are located. It changes everywhere', Taylor said. Nevertheless the global impact was rather puzzling, since sulphates by themselves don't cross the planet, but northern hemisphere pollution, apparently, can change the atmosphere worldwide.

Building chimneys taller to disseminate the pollution away from the ground, say Columbia University scientists, makes it worse.

Satellites offer the best way of detecting such effects, but it is not always easy to do this over land surfaces. Measurements in the late 1980s of the impact of smoke from ships' funnels have provided clear evidence that particulates increase the reflectivity of clouds and hence are more likely to induce cooling.[30]

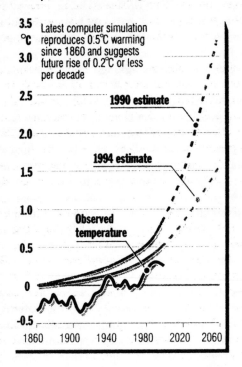

Climate scientists are now paying more attention to the cooling characteristics of sulphate aerosols.

Source: Sunday Times

The sulphate cooling theory is now almost an orthodoxy. A 3-month computer simulation by the Met Office in Bracknell, Berkshire, found that some 50 per cent of the warming earlier predicted had vanished.[31] David Carson, director of the Hadley Centre for Climate Prediction and Research, funded by the Met Office and the Dept of the Environment, said he now believed that global warming would be '0.2C or less' for the foreseeable future — difficult, he added, to distinghish from natural variability.

The Hole in the Sky

Ozone, however, has added to the confusion. In the public's mind the notorious ozone hole 'lets in' solar radiation that is a contributary factor in the warming. In fact the ozone layer has nothing to do with the greenhouse effect, yet lengthy discussions of it are often included in greenhouse books. This is because of its interaction with other stratospheric, and apparently warming chemicals and the sun's rays. Yet it merely serves, in the last analysis, to advance aerosol cooling theories.

In reality one is talking about tiny concentrations (of about five parts of ozone to every million parts of other gases), and they exist at latitudes so high that little is, in truth, known about ozone by scientists. In recent years ultraviolet light reaching America seems to have fallen off. But this is largely, as we have seen, because of the increase in particulate matter, and partly because ozone is a pollutant in the troposphere (there are in fact two ozone layers; one at virtually surface level, causing pollution smog, and the other about 20 miles up).

Sherry Rowland, one of the key scientists who drew the world's attention to the ozone 'hole' in the 1980s, says the worst climatic effect would be to change weather patterns by, for example, disturbing the jet streams.[32] Tropospheric ozone will in principle add to the warming because it is an efficient absorber of IR radiation in the band from 8 to 10 micrometres. But anti-greenhouse campaigners seem more concerned about stratospheric ozone. John Austin of the Met Office and two other climatologists calculate that if atmospheric CO_2 doubles, say as predicted by the middle of the next century, virtually all the ozone in the lower atmosphere will be destroyed.[33] This is because greenhouse gases trap heat near the surface, thus robbing, in a sense, the upper atmosphere, or stratosphere, of warmth. A vicious circle is set up. A warming at surface level will make the top of the stratosphere cooler, which in turn tends to increase ozone because ozone-destroying reactions start to slow down at lower temperatures.

The spring temperature above the Antarctic is normally icily cold, but even more so during an atmospheric warming. Icy particles form in polar clouds, and reactions to these particles then release chlorine. And chlorine, we are told frequently, is the main 'destroyer' of ozone via photochemical reactions. CFCs, by releasing chlorine, are also said to act in a similar way. However, ozone is as difficult to destroy as a Lego-built toy truck. The sooner the truck is pulled apart the quicker another one can be built.

Ozone (tri-atomic oxygen) in the stratosphere is a self-regulating mechanism, being formed by the continual breaking up of oxygen, which then acts as a barrier to further UV radiation hitting the Earth's surface. This barrier varies from season to season and from day to day because of the intensity of sunlight. The ozone layer is thickest over the north and south polar regions during the summers, when there is daylight arriving continually at an oblique angle and striking ozone and oxygen molecules.

The reason the air over the north pole does not get as cold as that of the Antarctic is largely due to the distribution of land masses, which differs in both hemispheres (there is more land north of the equator). The Antarctic also suffers from the blistering cold whipped up by the circumpolar vortex, which howls across the unobstructed southern ocean.

Many climate commentators sometimes seem to forget that the Earth is a rotating planet. Around the globe the surface warms and cools alarmingly as day follows night. Studies from the Lamont-Doherty Geological Observatory at Columbia University, New York, reported that warming in the southern hemisphere occurs only in winter night-times.[34] This is because during the day increasing cloud cover acts as a screen, but a ground heat-trap during the night. This change in day/night balance also implies that this kind of aerosol-cooling is altering the vital heat-and-energy physics of the atmosphere, the land and the oceans.[35] The thermal gradient comes into operation — one of the factors that puzzle scientists in the advent of an atmospheric warming. For example if a cooling is concentrated in mid to northern regions it increases the temperature differences between these regions and the tropics.[36]

Even the belief that CFCs warm the atmosphere is under attack for similar reasons. Haroun Tazieff, a vulcanologist and environmental adviser to the French Government, says CFCs must play a minor role because only 7,500 tons of chlorine are released from the breakdown of CFCs annually, against 660m tons from the evaporation of seawater and volcanoes[38]. Tazieff doesn't believe in the ozone scare, saying it is a natural phenomenon that occurs in the Antarctic spring each year between October and December: 'In Europe I think I am the only person to refute it, and I have never been officially contradicted, neither by ecologists nor by scientists'. In reply to an assertion by Friends of the Earth that mid-latitude ozone was being depleted by 3 per cent a decade, he said: 'It is absolutely untrue'.[38]

Many believe it is volcanoes that thin out the ozone layer simply by cooling the globe with a dust veil. The best known example was the 1916 eruption in Indonesia of Tambora. It brought about what has commonly been known as the 'year without a summer', when parts of North America were hit by snowstorms in the summer. And in 1991 aerosols from Mount Hudson in Chile destroyed ozone below 13 km. During September of that year about half of the ozone at altitudes of between 9 and 13 km were destroyed. Between December 1991 and March 1992 the amount of pollution in the stratosphere above Thule, Greenland, was monitored from the ground by balloons launched by the Danish Meteorological Institute. Its scientists found Pinatubu aerosols had penetrated the Arctic stratosphere in thin layers below 16km.[39] According to the Nasa-Ames Research Centre five months after the Pintabu eruption ozone in the lower stratosphere was reduced by 30 per cent[40]. Ozone destruction in the tropics is more a function of photochemical reactions than in polar regions. It changes seasonally and diurnally, but more ozone is created than destroyed[41].

Of more importance was Pinatubo's physical effects. It pumped 20 million tonnes of sulphur dioxide into the stratosphere, reducing solar penetration by around 4 watts per square metre, exceeding the magnitude of the warming[42], cooling the world by about 0.035C. Climatologist James Hansen of Nasa says Mount Pinatubo 'may be the largest global climate perturbator of the century'.[43] It just proves, he says, how sensitive the atmosphere is to small changes in its heat balance. Further, the volcano's sulphuric acid droplets may have slowed the growth of bacteria on land. Since bacteria eat dead plant matter and return CO^2 to the atmosphere, this could also keep more of the gas out of the atmosphere. Carbon dioxide, the main agent of recent warming, now appeared to have too many adversaries to do its job effectively.

FOOTNOTE REFERENCES:

Chapter Three: The Cool Skies
1. New Scientist, 17/10/92.
2. Science (US), vol 263, p1587, April 1994.
3. New Scientist, 5/11/94.
4. New Scientist, op cit, 17/10/92.

5. New Scientist, letters, 3/12/94.
6. *Hothouse Earth*, John Gribbin, Bantam Press, 1990, p132
7. Ibid, p130.
8. Discover, (US), July 1993.
9. Spectrochimica Acta, October 1994.
10. Discover (US), November 1992.
11. New Scientist, 9/7/94.
12. Ibid.
13. Ibid.
14. Ibid.
15. Geophysical Research Letters, July 1994.
16. New Scientist, ibid, 9/7/94.
17. Ibid.
18. Ibid.
19. Discover, (U.S.) ibid, July 1993.
20. New Scientist, ibid, 9/7/94.
21. Discover, ibid, July, 1993.
22. Ibid.
23. New Scientist, 29/10/94.
24. Ibid.
25. New Scientist, ibid, 9/7/94.
26. Discover, ibid, July, 1993.
27. Earth Magazine, (U.S.) December 1994.
28. Sunday Times, Hilary Lawson, 12/8/90.
29. Sunday Times, 24/9/94.
30. The Times, W. J. Burroughs, 4/3/88.
31. Sunday Times, ibid, 24/9/94.
32. *Turning up the Heat*, Fred Pearce, Bodley Head, 1989, p24.
33. New Scientist, 28/11/92.
34. New Scientist, 28/1/95.
35. New Scientist, ibid, 9/7/94.
36. Ibid.
37. Ibid.
38. The Times, 10/10/94.
39. Ibid.
40. New Scientist, 28/11/92.
41. New Scientist, John Gribbin, 2/1/93.
42. Ibid.
43. New Scientist, Fred Pearce, 19/6/93.
44. Ibid.

Chapter Four:

THE OCEAN THERMOSTAT

Those who worry overmuch about the greenhouse effect often forget that the oceans govern the climate. The sun sets it in motion, of course, but the oceans regulate it and dominate the climate and ultimately determine its every change of direction, down to the last half a degree Centigrade, and keep it that way — cooler or warmer — for literally hundreds of years.

Climatologists frequently remind us that the oceans function like a gigantic storage heater, absorbing five times as much solar heat as dry land, slowly discharging it as the globe rotates and orbits the sun. I will have more to say about the sun in Chapter Seven, but for the moment I will play devil's advocate and say that in theory there is no need to introduce the sun at all.

The main stress forces on the planet arise from its own violent internal dynamics. More important, they are responsible for its *thermal* structure. Pressure and temperature — i.e. physical rather than chemical features — explain most of the variations of the Earth's crust, and below that the mantle. Scientists are only now gaining a rough idea of just how hot is the Inner Earth. Geophysicists at the University of California, Berkeley, in the late 80s achieved temperatures of about 6,300C by subjecting core-like iron oxides to massive pressures equivalent to what most scientists speculate occur at the boundary between the inner and outer cores of the planet.[1]

If indeed the world is 'getting hotter', then some powerful energy force is bringing it about. Physicists rank energy-forms from the terrestrial to the cosmic: first we have mechanical energy, then thermal, electric, chemical, and finally energy-in-matter. It is from the last, in a real sense, that all other energies are derived, as illustrated in Einstein's famous equation $E = Mc^2$ (which incidentally, is the only mathematical equation you are likely to see published in a newspaper). For our purposes it is enough to say that energy is created when differently charged atomic particles collide with each other. Chemical energy is released after being stored in the linkage of atoms to molecules, and compounds vary in their energy richness. It was when these bonds are broken that energy escapes as heat or light.

Chemical reactions rearrange molecules, using the same principle of a descent into chaos, simply by dispersing energy. Food, as a product of a stream of never-ending corruption, by providing energy for animals, is a kind of chemical heat engine. The atmosphere will generate both mechanical energy and chemical energy arising from its chemical structure, but clearly the former will vastly out-perform the latter. But not only is external heat (i.e. solar radiation) outside the physicist's list of terrestrial energy forms, heat, whether endogenous or external, is like a bad currency. Both electrical and mechanical energy can also create thermal energy, but thermal energy can seldom be exchanged for any other kind of energy, unlike all other forms. Energy can be put to work only if there is not an even flow in its concentration. In other words heat arises after conversion from energy originating at a higher point of concentration.

The energy-heat equation works much more effectively the other way. The oceans contain vast amounts of kinetic terrestrial energy. They are like the armatures of a giant dynamo, powered by over a billion megawatts every day — using the dynamics housed in over 368 million square kilometres of liquid chemicals that cover some 70 per cent of the Earth's surface. This is why the Earth appears blue from outer space. Indeed the blueness, together with the white streaks of the clouds, draws attention to the hydrospheric, and hence heat-controlling, characteristics of the Earth. For example a physicist would know that transforming a gram of water into vapour takes 580 calories, which is then released when the vapour condenses perhaps hundreds of miles away onto land masses.

As in all dynamos, physical events take their time to work up to their maximum capacity, and take a long time to slow down. Deep dragging currents take literally tens of years to transfer trillions of watts of heat from one side of the world to the other, first from south to north, and then laterally across all the world's interconnecting oceans. Annually this represents the equivalent of the output of one million power stations.[3]

But bear in mind that the whole is complicated by Earth's axial spin. Since the 19th century the ocean currents have been thoroughly charted. It seems they move in large clockwise circles in the oceans of the northern hemisphere, and counter-clockwise in the oceans below the equator. This flowing and reversing phenomenon is a troublesome syndrome responsible for much of the world's more baffling climate and weather anomalies, but its causes can be explained simply enough. When anything moves across the turning Earth, say an artillery shell, it is deflected to one side. When the atmosphere is involved the phenomenon is known as the Coriolis Effect, and is described in what are known as the Navier-Stokes equations. The ocean currents are forced to curve along a great series of circles known as gyres.

The well-known Gulf Stream is one such gyre, and is the westernmost component of the northern clockwise current. The Gulf Stream transports vast amounts of heat hundreds of miles, and is driven by a clash of temperatures from the hot equator to the cool poles. Warmth is imparted to the adjacent land because of winds blowing across the warm waters.

The discovery of the Gulf Stream as early as the mid 19th century was by an American naval officer named Matthew Fontaine Maury. He gave it a description that has become a classic remark in oceanography: 'There is a river in the ocean'.[4] This 'river' is 50 miles wide at the start, and nearly half a mile deep, and pumps a thousand times as much water each second as does the Mississippi. Even at slow speeds it can shift seawater northeastwards off New York City at about 45 million tons per second. By the time it veers towards Europe it bypasses Labrador in Canada leaving it bleak and desolate, and yet it is a landmass that is at virtually the same latitude as Britain.

In fact this portion of the world, the Gulf Stream-Labrador current boundary, affects the energy supply of the atmosphere, and steers some tricky weather systems towards Europe from time to time. Gulf Stream or no, Britain is not completely immune to the vagaries of her northerly altitude. Nevertheless the Gulf Stream reaches as far as the eastern part of the Norwegian Sea, and the average temperature over the year at Tromso, Norway, shows the air temperature, even in winter, can be about 14C above what one would expect. On the other hand the mean annual temperature at Quito, on the coast of Ecuador, is nearly 14C below the average for the latitude because the coast is perpetually cloudy, and this in turn arises from the continual upwelling of cold water.

A similar, but larger, counter-clockwise current swirls along the rim of the Pacific Ocean, moving northwards from the Antarctic up the western coast of South America as far as Peru — hence it is known as the Peru Current (See Chapter Five). The configuration of the Peruvian coastline combines with the Coriolis effect to send this current westward. Some of this water drifts as far west as the Indian Ocean, with much else of it moving southward to Australia, and then eastward again. Much of this movement has a heat-equalizing effect, but the various currents and gyres imply there is still a lot of uneven ocean temperature. The largest ocean current in the world is that circulating around the continent of Antarctica from west to east across unbroken waters, where it is clear of the continents of South America, Africa and Australia. It moves nearly 100 tons of water eastward each second.

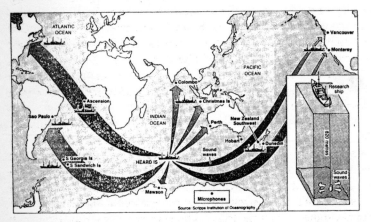

Taking a sounding: Loudspeakers lowered from a US navy ship (above, insert), moored off Heard Island emit intense sounds which can travel to each of the world's interconnecting oceans. Microphones in the sea at various bases can pick up the sounds. The speed of the acoustic signal varies according to the temperature of the water.

Source: The Times

However, northern European temperatures can be greatly affected by the fluctuating nature of the pack ice floes around the Arctic Ocean. It has been found from studies in the North Atlantic-European sector that in years when the pack-ice limit is unusually far south the eastward-travelling, rain-giving cyclonic depressions in summer and autumn are more prevalent more further south than usual. Often a poor European summer can simply be due to this.

Indeed the climatological knock-on effect of solar radiation is disproportionately huge. For example Earth's temperature surges and subsides alarmingly,[5] making land temperatures throughout the year vary by as much as 80C, whereas those of oceans at the same latitude vary only by 10C. Yet the minuscule skin of the Earth cannot really be compared with the depths of the oceans. More heat is trapped in the topmost layers (confined to the top 200 or 300 feet) than in the entire troposphere with which it is in constant interaction.

In fact the global warming argument can be stood on its head. Quite a small change in ocean surface temperatures can have marked effects on the atmosphere, especially in the tropics, where the heating of the oceans by the sun quickly saturates the air with water vapour, which condenses with such overpowering energy that violent hurricanes are unleashed — where, in effect, convection heat is turned into kinetic energy.[6] The currents and flows within the oceans play a more important role in mimicking the *behaviour* of the atmosphere above rather than absorbing heat from it (although it can absorb energy from it).

Interactions between atmosphere and ocean also operate at different rates, producing lag effects, which long-range weather forecasters can never neglect. If, for example, the trade winds in the Atlantic close to the equator are blowing harder than usual, more water is forced in the direction of the Caribbean and the Gulf of Mexico, producing a stronger flow and greater warmth in the Gulf Stream some six months later.[7] The currents mix warm water with cold down to the lower depths (to what is known as the thermocline) over a period of hundreds of years.[8] Warmer atmospheric conditions seem to play merely a walk-on role.

The Eye in the Sky

One important aspect of climatological research — now a multi-million dollar industry — is to find out whether the oceans are warming perceptibly. Since the middle of the 19th century scientists have rigorously analyzed the motions and contents of the sea. They have probed its raw, surging energy, and have charted its vast horizontal surface, its submarine currents and powerful upthrusting vertical movements. Oceanographers have mapped the seabed and its shelves, slopes and ridges — indeed, the discovery of a ridge-rift system through all ocean basins first led to the new concept of seafloor spreading.

But to determine whether the trillions of tons of dense, saline, cold water is becoming less cold is a very tall order. Checking out the ocean depths was a task hitherto confined to unglamorous cruises on research ships, often in inhospitable conditions. Measuring seawater temperature is highly dependent upon density and salinity. In the '50s and '60s observations were confined to visual data, but in the '70s, for the first time, electrical wave measuring devices provided fresh data.

Modern techniques are now innovative, and ingenious examples of applied physics are in use which are an impressive improvement on earlier techniques. Critics, however, might question whether one kind of physics can help understand another — small man-made bleeping and screeching devices are used to try to understand immense forces brought about by pressure, motion and temperature. The task seems unequal. In principle instruments can, but in practice cannot, cope with all the complexities of ocean physics, or adequately interpret what is happening, and I will explain why a little later.

First, to set the sceptical tone, I must return to the subject of the last two chapters — *carbon*. Do we actually know, and have we a way of finding out, how much carbon is in the oceans? There is still much speculation on the subject. In their book *The Greenhouse Effect*, for example, Stewart Boyle and John Ardill said that 'only half the extra carbon dioxide added by human activity remains in the atmosphere . . . Some of it may have been taken away by vegetation; most of it has probably been absorbed by the oceans . . . scientists are unsure how much extra capacity the oceans have for absorbing carbon dioxide. Some suspect they may have just about reached their limit. Others think [otherwise]. There is evidence that the carbon content of the oceans has varied over long periods of time . . . ' and so on.[9]

Some scientists have said that of 'seven gigatonnes' of carbon dioxide, 'some four gigatonnes were being sunk in the oceans'.[10] One journalist writing for the popular science magazine *Focus* in August 1994 said perhaps 98 per cent of the Earth's carbon dioxide remains 'quietly dissolved in the oceans'. There are, says Peter Brewer, a senior scientist at the Woods Hole Oceanographic Institute, Mass, an estimated 1600 billion tonnes of dissolved organic carbon in the ocean, more than all the carbon stored in trees, grass and other plants.[11] Brewer complained that until the late '80s none of this carbon was included in models of the Earth's carbon cycles.

With this last point in mind let us turn to the subject of satellite surveillance. Essentially, there are two types of satellite: geostationary and polar-orbiting. The former are in 'geo' orbit about 20,000 miles above the equator, and always view the Earth from the same position. Polar satellites follow a north-south orbit, but both types depend on the Earth orbiting below them which they view on successive days in strips.[12]

Geostationary satellites sit high above a fixed point on the Earth, keeping a continual eye on shifting weather patterns; they can determine and track the severity of storms and their paths,[13] and send down images every half hour. But polar satellites have better resolutions since they are always moving relative to the Earth's surface. Even so, although up to two million square kilometres of ocean can be covered over several months,[14] only a minute portion of the 1,400 million cubic kilometres of the ocean can be measured. In some ways straight-forward optical imaging (in effect, picture-taking) can reveal more than other more subtle electromagnetic procedures. Cameras on board can perceive different shades of blue ocean water. This is because sunlight is absorbed at different spectral frequencies depending on what is dissolved in the upper layers.

Green waters are tinted by chlorophyll and other plant pigments derived from plankton. And the deep azure of some tropical waters is due to their having little life in them, in contrast to the grey-greens of coastal areas. Most of the colour-change is due to the decay of tiny unicellular creatures known as phytoplankton. It is a geochemical process: plankton take up inorganic carbon from the upper waters of the ocean; some of the carbon sinks to the seabed, and yet more is taken in from the CO_2 in the atmosphere, in a cyclical fashion, by other plankton. Sometimes this cycle is referred to as a 'biological pump'.[15]

One important recent discovery was the degree of plankton 'bloom' at high latitudes in the spring.[16] In the late '80s the Joint Global Ocean Flux (JGOFS) expedition, involving many European research ships, was devoted to studying biological and chemical processes that recycle carbon. Turbulent winter weather in temperate zones stirs up the upper ocean, bringing rich nutrients to the surface. Then, as the ocean warms, and stabilises as the seasons change, strengthening sunlight creates an explosive growth of nutrient-consuming phytoplankton. This massive spring bloom, bigger than anyone suspected, soon sinks rapidly to the depths of the oceans, initiating a cycle lasting probably one thousand years or so before they reach anywhere near the surface again.[17]

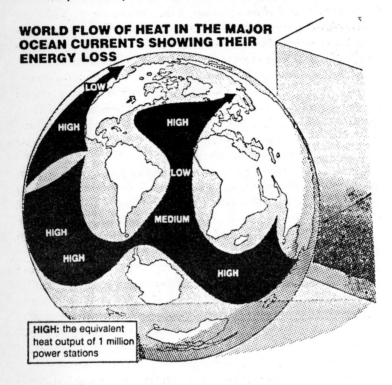

WORLD FLOW OF HEAT IN THE MAJOR OCEAN CURRENTS SHOWING THEIR ENERGY LOSS

LOW

HIGH   HIGH

LOW

MEDIUM

HIGH

HIGH

HIGH

HIGH: the equivalent heat output of 1 million power stations

The oceans dominate the climate because they contain vast amounts of terrestrial energy, equivalent to over a billion megawatts every day.

This phenomenon is a vital clue to what is happening to Earth's store of carbon dioxide, because in addition to the phytoplankton the Ocean Flux researchers noted vast numbers of zooplankton consuming the microscopic plants. Hence the plankton bloom continues to expand because the zooplankton also recycle nutrients that the phytoplankton need, thus prolonging the bloom which mops up still more carbon.[18]

There is a clearly defined line of colour along the equator in the Pacific Ocean. The Earth's rotation obliges easterly equatorial winds to diverge. Warm, shallow currents flow westwards across the Indian Ocean and round the Horn, up the eastern Atlantic to Britain, and then turn down the western Atlantic and southern Atlantic in a deep, cool and salty sweep to pass back eastwards by way of the Southern Ocean (which itself plays a big part in the global carbon cycle).

How, in the meantime, does oceanic organic life relate to the greenhouse effect? Some 100 climate modellers were spurred to meet in June 1990 at the Royal Society in London to 'discuss how the warming caused by the greenhouse effect would lead to a change in the upper layers of the oceans, disturbing the way plankton grows'.[19] A warm sea, they believed, may diminish its ability to absorb carbon dioxide, thus leading to a yet warmer atmosphere, and so on, creating a viscious circle, with more and more greenhouse ingredients coming into the equation.[20]

But were the Royal Society scientists aware of the findings of the Gobal Flux team published a year earlier? A warm sea, according to their findings, would surely stimulate plankton growth further, and CO_2, as we have seen, would be 'consumed' in ever greater quantities.

How, we might go on to ask, exactly does CO_2 warm the ocean surface? NCAR scientists suggest it produces a warming in the overlying atmosphere which may alter its 'vertical structure' and 'stability'. The main reason for the difference in heat retention properties, according to Andrew Heymsfield and Larry Miloshevich, is that whereas solar heating warms the surface of the ocean, energy absorbed by CO_2 is trapped in the body of the atmosphere and will not stimulate moist convection in the same way.[21] This in turn could produce the all-important feedbacks.[22] Two scientists at the Scripps Institute analyzed data obtained by satellite, and looked at cloud cover changes during an important Pacific warm current reversal, (known as El Nino — see Chapter Five). They too declared that carbon dioxide greenhouse warming operates in a 'different way' from the cycle that causes the ocean warming.[23]

I will have more to say about ocean warming, El Nino and CO_2 in the next chapter. In the meantime it is worth bearing in mind the most important self-correcting feedback known to climatic science: cloud creation out of moisture convection. As oceans warm, convection in the atmosphere increases, carrying moist air to the top of the troposphere and producing cumulonimbus clouds. These trap heat, but as convection gathers momentum icy crystals form thin cirrus clouds and deflect and scatter sunlight, preventing it from reaching the sea surface, so a new equilibrium is established.

Further, George Tselioudis at Nasa's Goddard Institute for Space Studies in New York has studied new satellite data from the International Satellite Cloud Climatology Project and found that variations in sunlight reflected by low clouds may offset a greenhouse warming,[24] although it was not possible to give a detailed scientific account of how this happened.

There is still no scientific consensus: perhaps thick clouds cool the atmosphere, while thin, high clouds act as greenhouse gases.[25] It could be, said Tselioudis in January 1994, that the higher the temperature the more cloud reflectivity drops. But how much so depended on seasons, latitude and whether the clouds were over the land or sea.[26] On the other hand Tim Palmer of the European Centre said that negative feedbacks associated with the day-to-day weather may offset positive climatic feedbacks, making the climate more self-regulatory than we think.[27] None of this, incidentally, has yet been satisfactorily demonstrated in computer models.

Taking a Sounding

Clearly, then, organic oceanic life is sending out its own mixed signals to researchers. There are simply too many complicated feedbacks involved to understand properly what is happening. The difficulty arises from the physics of the deep ocean itself. Oceanic flows provide important clues as to the *nature* of organic life, but they provide few clues as to its cause or origin. Our knowledge is limited to distinguishing colour from plant pigments and that from other organic sources. As a result we have a cause and effect riddle caused by the 'plankton multiplier', as John Woods of the Natural Environmental Research Council put it.[28] Predicting the behaviour of phytoplankton is highly dependent upon the accuracy of weather forecasting techniques in relation to seasonal changes.

More important, it is virtually impossible, except over a very long time-span, for the variously heated layers of the ocean to mix in with each other. This sometimes gives an impression that the seas are stratified rather like rock beneath the seabed, and the atmosphere above the waves.

Let us look, first, at how this image plays havoc with a technique known as *acoustic tomography*. Oceanographers first became aware of the possibilities of bouncing sound waves off the sea bed after the First World War when German Navy scientists attempted to clear harbours of mines using primitive sonars.[29] During the immediate postwar period, theoretical studies enabled oceanographers to translate water temperature and salinity data into water-flow data (speed and direction). The currents could not be observed directly until the 1950s when improved marker buoys could identify deepwater currents and a modern echo sounder could provide a continuous silhouette of the sea floor.[30]

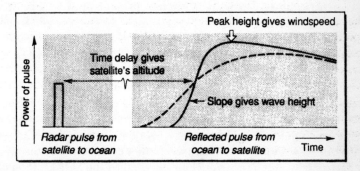

Radar reflections bouncing back to orbiting satellites give an indication of wind speed which in turn may be the result of atmospheric temperature changes.

Source: New Scientist

But modern variations of this technique do more than measure inanimate objects. Sound, like light, can be refracted, so that varying densities of water reflect sound waves differently. The speed of sound increases as the temperatures and pressure of the water rises. Acoustic tomography is theoretically similar to medical tomography, where X-rays map the density variations of the body by passing the rays through many different paths and then mathematically recombining them to form a 3-D image. Carl Wunsch of the MIT in 1979 reasoned that by transmitting acoustic signals over many paths — up, down, obliquely and sideways, and criss-crossing deep within the ocean — scientists could deduce the properties of the ocean's interior — its varying density, temperature and salinity.

According to Walter Munk, an acoustic experimenter at the Scripps Institute, the sound-time taken could increase by as much as a quarter of a second as a presumed warming takes place, and a rise of even a fraction of a degree should be detectable. John Spiesberger of the Woods Hole Institute, who is said to have pioneered acoustic tomography, proposes an $11-million space-age technology that would have 12 transmitters moored to the Pacific Ocean floor bleeping signals to the surface from receivers which could be off-loaded from vessels at determined points.[31] In 1991 sounds as loud as a jet taking off were made from loudspeakers lowered from a US navy ship, Corey Chouest, moored off the appropriately named Heard Island situated between Australia and Antarctica. This volcanic land mass, covered in ice, was chosen because there are direct 'paths' along which sound can travel to each one of the world's five interconnecting oceans.[32] Microphones in the sea at various bases in the Pacific near Australia could in theory pick up the sounds.

The drawback is that below about 1,000 feet the increased pressure offsets the decreased temperature, so the speed starts to pick up again.[33] Even though such variations amount to only about 3 per cent of the average speed of sound in water, about 1,500 metres per second, they have profound effects. Sound waves above the depth of a minimum speed are refracted downwards. The net effect is to trap sound waves in a 'duct', where they cycle vertically through a horizontal column of water. This duct, called the SOFAR (Sound Fixing and Ranging) channel, has a minimum depth speed which is greatest at equatorial latitudes. At high latitudes, where solar heating is neglible, the speed of sound is determined almost completely by pressure. The waves are reflected upward only, bounce off the surface, scatter and lose energy, and with it the instrumental accuracy that scientists prefer if they are to make proper calculations about oceanic temperatures in specific regions. In a nutshell, it seems that some sounds cannot be dissipated upwards because of a boundary that exists between warm and colder currents. They cannot travel downwards either, because of increasing pressure. So the sound is funnelled along a virtual 'wave guide'.

Plumbing the Depths

Now we turn to the use of radar, which has problems of a different order. Radar operates on a longer wavelength than the light and heat spectrum, and is also quite different from the acoustic and colour spectrum. One-hundred metre images would show up on the instruments of radar-carrying satellites. Many of these images would be produced by ocean floor trenches, ridges and faults which influence Earth's gravitational field. In other words many satellites can map the deep sea-bed by recording the undulations of the sea's *surface*. Any change in the latter, then, would be attributed to currents, because the shape of the geoid, i.e. Earth's shape, would remain the same over millions of years of continental drift. Adjusting the microwave radiation to bounce back only from the sea surface can help scientists, by analyzing the return pulse, to work out wind speeds and wave heights.[34] One of the earliest ocean-monitoring satellites — named Seasat — was launched in 1978 by Nasa.

Increasingly convinced of the warming/oceanic link, scientists are devoting enormous financial resources to extending the range and scope of biosphere-monitoring techniques. They are deploying a vast array of electronic gadgets that are byproducts of the hard-pressed space and defence industries. The International Council of Scientific Unions initiated a concerted global effort in 1990, involving scientists from 50 nations,[35] as well as the World Met Organization. The 10-year project was called the World Ocean Circulation Experiment (WOCE), and is a plan to deploy NERC ships to take 24,000 profiles of temperature, salinity and selected chemicals. Involved in the WOCE enterprise is Aston University, the James Rennell Centre for Ocean Circulation at Southampton University, and the Mullard Space Science Laboratory, which is playing a crucial role as part of an initiative in Earth Observation Science, with teams probing the interaction between all the major land/ocean/ cryosphere/atmosphere interfaces. The British Antarctic Survey, in conjunction with America's National Oceanic and Atmospheric Agency (NOAA), is working on problems associated with atmosphere changes and climate cycles around the Southern Hemisphere.[36]

In 1991 the European Space Agency's first remote sensing satellite, ERS-1, was launched into near polar orbit by an Ariane-4 launcher as part of the WOCE exercise. The ERS-1 carries on board the innovative European Along-Track Scanning Radiometer (ATSR). The second of its type, the ERS-2, was launched in 1994.[37] And in 1999 an advanced ATSR is planned to be launched on the ESA's environmental satellite, ENVISAT-1.

The ATSR, described by David Llewellyn-Jones, one of the ATSRs team based at the Rutherford Appleton Laboratory in Oxford,[38] as 'a beautiful piece of technology and science', will measure sea surface temperatures to 0.25C with an infrared (IR) detector. It will also determine, almost as an afterthought, what amount of sunlight falls onto the oceans. The machine is said to be 10 times more sensitive than experiments to date, allowing unprecedented accuracy in working out oceanic heat balances.

Scientists at the Proudman Oceanographic Laboratory at Birkenhead, Merseyside, will be using their knowledge about tidal changes and sea levels to ensure that the satellite's altimeter is making accurate measurements,[39] and researchers at Edinburgh University are to do some of the computer development work.

There are several parallel operations taking place. The ALACE, (Autonomous Langrangian Circulation Explorer) is a 120 cm long tube with an overall density somewhat greater than surface seawater to ensure that it will sink down to 1,000 metres, where it reaches water of the same density. The device can then ascend when a battery-powered pump pushes oil into a membrane which then expands to increase the instrument's overall volume. The ALACE signals report back to Russ Davis, an oceanographer at the Scripps Institute. By the turn of the century, Professor Davis hopes to have an armada of 1,000 ALACEs. These will supposedly give him the first detailed maps of ocean currents, which will also be part of the WOCE.[40] 'Operation Vivaldi', which started in 1991, has also successfully studied the effect of salinity on the 'conveyor belt' of giant movements of water in the interconnecting world oceans.

The WOCE has recently acquired yet another bureaucratic tier, the Global Ocean Observing System, a UN-backed initiative to set up a network of satellites, automatic buoys and unmanned deep-sea probes.[41] It is hoped this network will be in place by 2007, and will offer a massive array of up-to-the-minute knowledge about the pathways of the currents and their temperatures, and wave speeds and size.[42] Miniature submarines also play a useful role, such as the two pilotless carbon-fibre subs designed by the Institute of Oceanographic Sciences, part of the NERC. Robot subs have also investigated the feedbacks involved with melting ice, thickness of ice, and currents beneath the ice caps.

The technical and theoretical problems nevertheless remain formidable. WOCE does not pay sufficient attention to the polar regions, although the British Antarctic Survey part of the WOCE enterprise is supposed to redress the balance here. The polar regions are important because key feedback processes predominate. As Garth Rees of the Scott Polar Research Institute in Cambridge points out, 'the polar ice acts as a valuable marker for global climate change'.[43] Factors such as the Gulf Stream and the North Atlantic Drift, i.e. all those factors which feature as explanations for 'climatic flips' (see the following chapter), are poorly represented in the computer models. This probably explains why research on the transportation of water in the atmosphere had models that were 'off by a factor of two', Richard Lindzen points out.[44] Current models exaggerate moisturization factors, especially in the upper troposphere where humidity, as many models proved, often traps most of the Earth's heat.

Climatologists will often use the word inertia to explain why the oceans act in such excruciatingly slow motion — taking up to 100 years or so to affect any noticeable change. Hence the need for WOCE. Yet the matter is fraught with real scientific handicaps that scientists would be better off not mentioning! Dr Woods says he is still not certain whether we can predict ocean-based climatic changes.[45] The WOCE exercise is designed to whittle down some of this uncertainty and to improve our oceanic knowledge in a single crash programme that eliminates some of the unsystematized and dispersed exercises of earlier decades.

Choppy Seas Reveal Much

A major role of satellites is to check up on what the surface of the ocean is doing. Currents change the height of sea surfaces, with warmer currents, such as the Gulf Stream, making noticeable differences. Massive eddies whirl away from the sides of major currents, dissipating thermal energy. But checking wave height, like monitoring the changing colour of the oceans, is only half of the task. Scientists are left with the conceptual problem of translating wave height (rather than sea level rise) into proven warming trends. The conventional wisdom is that a warming alters the thermal gradient, as we saw earlier.

But what do eddies, waves, mini-waves and sea currents imply? High winds roughen the surface of waves, and there is a time delay with the peaks and troughs of constantly flowing oceans. The return radar pulse — coming as much as 20 times a second — will be more randomly scattered. The stronger the signal, the lower the wind speed.[45] Much of this can be scientifically hazardous, as Andrew Lorenz, head of Data-assimilation Research at the Met Office, points out.[46]

It is only ripples, rather than the wind-speed, that are being measured, which means that a lot of mathematical processing, and hence more abstraction, has to be done. For example officials at the NERC have detected the north-east Atlantic getting rougher since the 1960s.[47] Wave heights were said to have increased by 25 per cent, with the biggest growing from 12 metres to 18 metres during the last 25 years.[48] Yet David Carter, one of the Institute of Oceanographic Sciences team which made the discovery, said that he did not know the reason for the observed increase in wave height. Some British scientists are checking to see if there are any telltale increases in storm activity in line with many greenhouse predictions, such as the one in late January 1990. Higher waves should mean stronger winds, yet this was not correlated with any other scientific findings.[55] But what do stronger winds themselves mean? Significantly Carter said it was too soon to blame wave height changes on the greenhouse effect: it could be just a statistical 'blip'.

This concept, the idea of the 'blip' — the technical or geophysical freak occurrence — in fact gets to the nub of the issue. In truth Seasat, ERS, and the other satellite instruments are part of the weather-forecasting world, where 'blips' happen all the time. Seasat scanning is characteristic of a growing trend in climatology where meteorological philosophy and hardware are used. This is partly because climatologists are agents of society, and the importance of the ocean as an applied field of study meets the pressing needs of that society. Researchers have to go along with the current concerns and anxieties of government-funded scientific institutions as priorities and concerns change, and as technology changes.

Since the late 1960s, after the wrecking of the tanker Torrey Canyon, new pollution perils have emerged, and the NOAA has repeatedly reported, from its satellite observations, the extent of pollution along the eastern coast of the US. To take another example, the Topex Poseidon satellite bounces radar beams off the sea, and as it passes 1,300 km over the English Channel its height will be measured by the Herstmonceau Castle in East Sussex. This laser checks the accuracy of a Franco-American satellite which lifted off on an Ariane rocket from the ESA's launch-pad in French Guinea in August 1992.[56] Yet the information gathered by Topex has been used to point to the safest place to dispose of hazardous waste, to help predict when icebergs cross shipping lanes, and to improve the fuel efficiency of ships by routing them along the fastest currents.

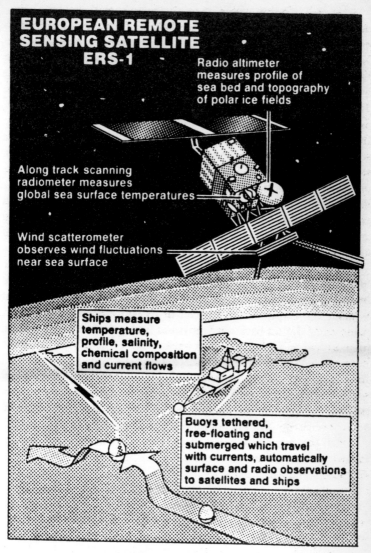

The ERS satellite is part of the World Ocean Circulation Experiment, one of several multi-million pound Earth-observing projects.

Source: The Times

However, like the climate modellers themselves, long-range forecasts can only be presented partially, and the difficulty of cross-referencing the various finely-tuned electronic signals can get overwhelmed by the oceanic 'noise' — the huge blurring qualities of the mass of the ocean. Nick Flemming, a senior scientist at the Institution of Oceanographic Sciences in Wormley, Surrey, said the barrier to long-term forecasts is imperfect ocean measurements, with forecasts still dependent on atmospheric measurements.

Another problem is that for statistical 'blips' to be ruled out experiments will need to be conducted over tens, if not hundreds, of years. Furthermore the clouds issue has been seriously neglected. This returns us to the subject of plankton. Powerful oceanic feedbacks have long been recognized, such as the sulphates which are to be found in dimethyl sulphide (DMS), the gas emitted by microscopic ocean-going creatures. DMS is often a favourite of climate and Gaia theorists who like to point to its remarkable countervailing properties. Indeed, if there is one organic phenomenon on Earth that could counteract the greenhouse warming alone, it would be phytoplankton. These tiny specks of organic life can create the nuclei of moisture droplets in clouds, which then deflect back solar radiation, and start to cool the surface that was originally beginning to warm.

In the meantime many more 'eye in the sky' projects are planned by the Americans. During the next ten years Nasa hopes to spend over $40 billion to launch satellites containing dozens of instruments for the Earth Observing System (EOS), the centrepiece of Mission to Planet Earth. Not only ocean warming and deforestation will be studied, but the satellites will be used by other nations and ground monitors to develop a baseline of information against which global change can be measured.[56]

FOOTNOTE REFERENCES:
Chapter Four: The Ocean Thermostat
1. *Fate of the Dinosaurs*, Antony Milne, Prism Press, 1991, p206.
2. *The Ocean*, Bantam/Britannica, 1978, p91.
3. New Scientist, 27/4/91.
4. *Asimov's Guide to Guide*, Penguin, 1987, p163.
5. Times Higher Education Supplement, 11/3/94.
6. Focus magazine, Aug/Sep 1994.
7. *The Ocean*, ibid, p88.

8. Focus, ibid; New Scientist, 27/7/91.
9. *The Greenhouse Effect*, Stewart Boyle & John Ardill, New English Library, 1989, p36.
10. Nature, vol 358, p723, September 1992.
11. New Scientist, 15/12/90.
12. New Scientist, 27/4/91.
13. Ibid.
14. New Scientist, Marlon Lewis, 7/10/89.
15. Ibid.
16. Ibid.
17. Ibid.
18. New Scientist, 15/12/90.
19. Daily Telegraph, 11/6/90.
20. Ibid.
21. New Scientist, 11/5/91.
22. Telegraph, op cit .
23. New Scientist, ibid, 11/5/91.
24. New Scientist, 22/1/94.
25. Times Higher Education Supplement, 11/3/94.
26. New Scientist, op cit, 22/1/94.
27. Telegraph, op cit, 11/6/90.
28. Ibid.
29. Guardian, 10/4/92.
30. *The Ocean*, op cit, p45.
31. Guardian, op cit.
32. The Times, 15/11/90.
33. Scientific American, October 1990.
34. New Scientist, op cit, 27/4/91.
35. Geographical Magazine, May 1992.
36. New Scientist 8/10/94.
37. New Scientist 27/4/91.
38. Ibid.
39. The Times, 18/8/92.
40. The Times, 26/8/92.
41. Time (US) 16/9/91.
42. Ibid.
43. New Scientist, 27/7/91.
44. See Chapter One.
45. New Scientist, op cit, 27/4/91.
46. Ibid.

47. Sunday Correspondent 28/1/90.
48. Ibid.
49. Ibid.
50. Daily Telegraph, 13/4/92.
51. Time magazine (US) 16/9/91.

Chapter Five:

THE REVERSING CLIMATE

The world's largest ocean, the Pacific, dominates our climate. One of the earliest discoveries about the atmosphere over the Pacific is that the winds flow regularly from west to east — the 'westerlies'. They are a constant factor, primed as they are by the sun. But they can be deflected, all things being equal, from a usually constant and calm path by the variations in solar radiation and by the geographic distribution of the land masses in the northern and southern hemisphere. But all things are not equal.

It is at this stage that we must introduce the jet streams, which have a very important explanatory role in the genesis of extreme or unusual weather. These exist about eight miles above sea level, and are the very embodiment of chaos theory — they follow no known laws and cannot be adequately predicted. The jet streams are great rivers of powerful winds that flow east to west — opposite to the normal Pacific air currents — in a zig-zagging, looping manner over the north and south polar regions. At times they drift further south, or develop a slight kink in the loops to bring massive depressions at surface — level dragging weather systems, both turbulent and balmy, hither and thither, slackening their own intensity and even reversing the direction of flow from west to east.

Unfortunately for science this reversing, meandering, looping nature of Earth's winds is inherent in Nature itself. For example the summer westerlies that bring the monsoons are connected to the Intertropical Convergence Zone (the ICZ) which straddles both halves of the globe depending on the seasons: if the ICZ drifts too far south in the summer it means that no rains will fall further north. Things could be even worse if the drifting gets dangerously out of control: some 40 years ago the westerlies used to bring mild winters to the northern US, but the pattern soon changed dramatically to bring harsh winters which continued right throughout the 1960s.[1]

But the important point is that the westerly pattern probably changed because they were, in a sense, following the lead of the meandering jet streams. In other words, if one wind system slipped, they all slipped in unison, playing havoc with rain-bearing clouds. Warm southern seas would reduce the strength of air convergence over Africa (for example), which in turn meant less upward motion, fewer rainstorms and so on. Some claim the 1987 storm in England was kick-started by the jet streams — by what one commentator called a 'jet-streak'.[2] Perhaps it was a left-over from Hurricane Floyd which had pumped the western Atlantic with a huge plume of warm air.[3]

Let me add that the zig-zagging patterns happen in slow motion — they do not imply a constant buffeting diet of erratic weather: quite the opposite. They are often responsible for stationary weather systems. Ridges of high pressure at surface level have a life of their own. They drift from west to east, and are often surrounded by low pressure systems — 'lows', in other words, surround 'highs'.

Great public annoyance is frequently expressed when the high pressure systems get stuck in one position for days and sometimes weeks on end — they become the infamous blocking highs (officially known as circumpolar anticyclones), and can bring on prolonged unbroken warm weather, or intensely dry and cold conditions in winter. A blocking high can effectively squeeze out milder weather patterns in a kind of pincer movement, and pump additional energy into the system from outlying areas to keep the anticyclone going.[4]

One theory says that northerners suffer from these blocking highs because of the distribution of oceans, land masses and mountain chains that are not the same as in the southern hemisphere. There is more land in the north than in the south, where the large oceans dominate (in fact, overall 70 per cent of the globe is covered by water, but it becomes 80 per cent in the southern hemisphere). The great 1976 hot spell in England was said to be due to a 'blocking high'. So was the American heatwave of 1988, when temperatures had exceeded 90F for 32 days since the beginning of June in New York, and topped 100F in 45 cities across the breadth of America. Temperatures in North Carolina stayed at above 110F for weeks, and much of the state's staple crops of tobacco, cotton and corn were ruined.[5]

Indeed, blocking anti-cyclones were blamed for droughts and floods in Europe and Russia in the mid 1970s, and the blizzards in the north-central zones of the US in January 1985 which took the lives of 80 people. Dozens more died in the north-east of the US a year later,[6] and in the freak blizzards and snow storms of January 1978, the early 1981 prolonged cold snap in Europe, America and the floods and drought in different parts of China a year later. By the 1970s there was already clear evidence that Sea Surface Temperatures (SSTs), those great currents and gyres, were of increasing importance in explaining most of the vagaries of climate, as were the jet streams.

Knowledge of SSTs owes its existence to a signal scientific advance, the genesis of which can be traced to the early years of this century. A Cambridge mathematician called Gilbert Walker was despatched to set up a Met Office in India in 1904. He focussed his attention on predicting monsoons, the failure of which could prove catastrophic to local inhabitants. While failing in this regard, he correctly perceived that large-scale oscillations in climate were related to the Pacific. He called it the Southern Oscillation, by which it is still known. He noted in particular that when pressure is high in the Pacific Ocean it tends to be low in the Indian Ocean from Africa to Australia. In both zones, he said, one noticed a fall in air temperatures and rainfall patterns are disturbed.[7]

Now we must identify the main player in the 1980s storms, the *El Nino* effect, which is intimately related to Walker's theory. El Nino is Spanish for small boy (or the Christ Child), because it usually occurs around Christmas time. The phenomenon is more correctly known as the El Nino/Southern Oscillation, or ENSO. This event begins with the warming of the Pacific surface waters, and a weakening of the trade winds, which normally push anti-clockwise round the so-called Peru Current, and starts to race back instead towards North and South America in a three-to-five-year cycle. The Pacific Ocean, we must remember, is the world's largest, and is the major Earth thermostat.

Important experiments carried out at the Max Planck Institute in Hamburg show you can cause intensive changes in surface conditions on quite a short timescale[8] when variably heated 'rivers' within the oceans, are artificially introduced. This implied that warm and cold patches of surface ocean water could set up distinct wind-flow patterns, which in turn act as a positive feedback to local sea and atmospheric circulation events.

Chris Folland and others at the Met Office recently fed SSTs into
the Met Office's sophisticated General Circulation Model (GCM)
(see Chapter Six) of the atmosphere which had software data that
included historic SST records going back to 1871.[9] With this record
the model generated figures for rainfall in the Sahel, and searched
for other correlations and anomalies. Folland found a clear
relationship between Sahel rain and the difference between average
ocean temperatures. He noticed that whenever the southern
hemisphere's oceans are warmer relative to those in the north, the
Sahel has a drought.

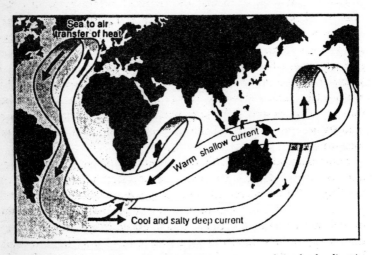

The broad sweep of the world's major ocean currents, their depth, direction,
temperature and salinity, largely determines the Earth's atmospheric temperature.

El Nino, however, more than any other SST, brought home to
weather men the troublesome nature of tropical ocean surface
anomalies. First applied to the coast of South America,
meteorologists soon began to relate it to the observed increase in
tropical oceanic temperatures. They guessed that its knock-on effect
was often heavy rainfall in the western Pacific and drenching rains
on the normally dry west coast of tropical South America.

The El Nino influence came to public prominence in 1982, when
rain patterns in the tropics and sub-tropics were altered to bring
more weather chaos. There were extreme droughts in Australia,
Indonesia, India, south-east Africa and Central America, and
thousands of lives were lost during a major African event. El Nino

has also been blamed for a famine in Bengal during the early part of the last war which killed millions, food riots in Zimbabwe, mudslides in Californian suburbs, and the end of the Peruvian anchovy and mackerel fisheries, not to mention the collapse of two ancient civilizations in Peru in 600 and 1100AD.[10] The Hadley Centre's sophisticated model of the global atmosphere showed an historic SST record going back to 1871. Its ripples even reach Europe. According to Robert Wilby of Loughborough University, it was responsible for bringing extra rain and snow to Britain, both recently and during the past century.[11]

However it does seem that the frequent failure to predict the onset of El Nino events, or to anticipate it when nothing happens, detracts from the forecasting prowess of climatologists. Forecasters seem to have little time to spot them and predict their paths. For example, journalists commented scathingly on the US National Weather Service forecast made in the autumn of 1990 for an El Nino-related cold winter in the eastern half of the US.[12] The forecasts were so low in the percentage of accuracy (often no more than 40 per cent chance of being right) that one might as well have tossed a coin.

Sometimes, instead, El Nino could be predicted accurately, but not its meteorological aftermath or its treacherous nature. A predicted 1991 event created a near crop failure in Brazil, with some parts of the poor north-east of the country receiving only a quarter of their normal rainfall.[13] It surged and ebbed frequently, but every time it appeared to have run its course it revived itself. Two years later it occurred unexpectedly again. Sea temperatures in the Pacific were 1.5C higher than usual, with 30C recorded where the International Date Line meets the equator. The late 1993 El Nino was longer, and matched historical events such as the phases that occurred between 1911 and 1941, 1939 and 1942. Only the NOAA model predicted this one because of the use of new sea and wind data collected from instruments moored across the Pacific as part of the International Tropical Ocean/Global Atmosphere (TOGA) research project.[14]

Soon all the world's aberrant weather, as before, was blamed on El Nino. In July 1994 a team of climatologists 'discovered' that the oscillation in SSTs caused by a new El Nino were strongly linked to more rainfall in Zimbabwe, floods in England, bush fires in Australia, and famine in Brazil.[15] In fact the team, led by Mark Cane at the Lamont-Doherty Institute in New York, was 'astonished' at how strong the effect seemed to be.[16]

Any causative explanation for ENSO events were disappointingly circular. It was said that sudden extra bursts of CO^2 into the air are due to El Nino creating periodic shifts in the direction of winds and ocean currents in the equatorial Pacific that in turn has global effects on the weather. Charles Keeling, a well known CO^2 trend watcher at the Scripps Institution (at La Jolla, California)[17] says variations in the level of the gas can make vegetation flourish and die back periodically. Again we have another complicated feedback: oceanic shifts ($\rightarrow$) storms and droughts ($\rightarrow$) density of vegetation ($\rightarrow$) more or less CO^2 and so on. It is sometimes suggested that CO^2 in turn has a further knock-on effect on swollen ocean levels. Stephen Zebiak and Mark Cane, both at Lamont, say that an El Nino event is always preceded by a swollen Pacific surface, because the thermocline (the barrier between warm and cold layers) fluctuates due to 'natural oceanic cycles'.[18]

How does El Nino fit into the greenhouse warming debate? Scientists say speculation along these lines is difficult to prove. 'It's a real chicken and egg problem', Nicholas Graham, an oceanographer at Scripps was reported recently as saying. 'If you have consistent El Ninos, then you end up producing a warmer world. So are more El Ninos causing global warming, or is global warming causing more El Ninos? I'd hesitate even to guess'.[19]

The Climatic 'Flip'
But talk of 'oscillations' and 'cycles' is rather like describing a wheel in motion. People would like to know, if the El Nino events are cyclical and periodic, what the terrestrial causative features might be. One could always come up with odd, spasmodic causes. Paul Handler of the University of Illinois once suggested that SST events are triggered by volcanic eruptions which occur between ten degrees south and 25 degrees north of the parallel. He cited El Chichon in Mexico in April 1982 and Nevade Del Ruize in Columbia in 1985.[20]

But no known terrestrial phenomenon could explain periodic or regular occurrences — only astronomical features could do that. By talking of 'natural variability' but denying the existence of natural cycles one is in danger of saying that climate is entirely the creature of *random* forces. Hubert Lamb, emeritus professor of meteorology at the CRU, is one of the few scientists who believe that bouts of stormy weather *are* periodic and to some extent predictable. He published in 1990 new data suggesting intense storms were actual repetitions of severe gales and flooding which affected the country in 1590, 1690, 1790 and 1890.[21] Most of these occurred during the 'Little Ice Age'.[22] Many ancient sea ports were literally obliterated from the map forever. A catastrophic storm in November 1703 killed nearly 10,000 people.

But periodicity implies predictability of a different order familiar to meteorologists. How could, for example, an anthropogenic warming, dating back to no more than 100 years, explain Lamb's 100-year cycle going back to the 16th century? Greenhouse-mongers will reply that past events and present ones are unrelated, although if climate reversals or other anomalies can persist long enough into the future the CO^2 argument is considerably weakened.

In the meantime what is frustrating, and engenders suspicion, is the fact that patterns of periodicity may break down because not even Lamb can say whether we are in for another stormy period. In other words we may be talking only of meteorological curiosities and statistical correlations and little else.

The term 'climatic flip' reflects the growth in status of catastrophe and chaos theory, after many years languishing in the realms of fringe science. It was somewhat inevitable, in the absence of an internally consistent theory, that chaos theory would come and fill the void. But to talk of current reversals, 'blips' and 'flips' is to introduce something quite beyond the climatologist's frame of reference. In theory a climatic flip could last long enough to turn Britain's equable climate into something approaching a Mediterranean paradise, or perhaps into a chilly wasteland like the Outer Hebrides. After all, the Outer Hebrides *is* Britain, and is only a few hundred miles from London. The slightest of gyre diversions is all that is needed to bridge the tiny geographical distance.

Further, both John Woods, director of Marine and Atmospheric Sciences at the NERC in Swindon,[23] and the IPCC in its report thought 'flips' would be constrained by the natural inertia of the oceans. The weather for a week or two ahead is dictated by the atmosphere, but over longer periods by the oceans. 'The atmosphere has no memory longer than about one month' said John Woods. But in truth the oceans have little memory either. 'Oceanic inertia' applies only to the total mass of the oceans. In other words nothing happens for hundreds of years until an unpredictable or peculiar rise in local SSTs has a knock-on effect.

Virulent storms could be part of a continuing and unknowable trend, perhaps always involving SSTs. The Met Office carried out an analysis of wind speed trends over the past 100 years and found the windiest occurring during the 1920s, and then in a later period from 1950 to 1964.[24] The period from the late 1960s to the present, it turned out, was among the least stormy since 1881.[25] In terms of one-off gales the mid 1930s and early 1960s were stormier than today.

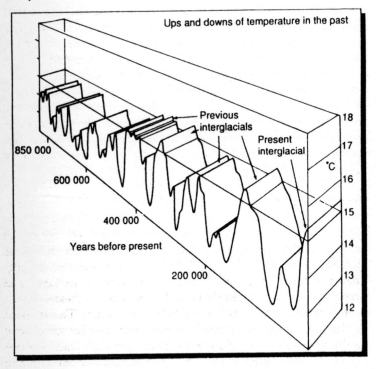

Earth's previous Ice Ages and interglacials were overwhelmingly the product of geophysical, orbital and astronomic factors.

Source: New Scientist

Climatology might well be heading for a crisis. Bill Burroughs, a physicist and frequent writer on climatic issues, said that he was 'throwing in the towel'. He wrote in *New Scientist* in 1994 that for many years he had sought to identify patterns in weather statistics that might provide the basis for seasonal forecasts. After all these years, he says, he may be 'barking up the wrong tree . . . it now seems that chaos reigns whenever it rains'.[26] He said he would normally believe in the efficacy of seasonal forecasts if it were not for the El Nino type of syndrome, which dramatically and unpredictably alters weather patterns. He says that he lost faith in weather prediction when the Met Office once said that climate variability this century has not, after all, much to do with observed alterations in SSTs and sea ice.

In other words even if one thought that El Nino events were the main meteorological player it didn't mean that any useful weather or storm predictions could be made either in the Pacific zones, Greenland or elsewhere. It is clear, said Burroughs, that historic annual figures make little difference to mid-latitude seasonal forecasts. Occasionally, he admitted, there was a hint that the history of blocking anticyclones matched up with forecasts of severe winters. But meteorology was not often to be classed as a reliable science — certainly predicting cold winters or warm summers was not possible. Such forecasts are 'built on and liable to be swept away by the ever changing chaotic eddies . . . '.[27]

Decoding the Ice Packs

These chaotic eddies are now beginning to worry climatologists largely because they challenge so many orthodoxies. The anthropogenic greenhouse theorists come off worst since their ideas are focussed forward upon a gradual long-term man-made warming. They pay scant attention to the Earth's historic ice ages and interglacials that have been the main determinant of climatic change in the past, and which clearly must do so in the future. This is not to say that ice age theories simplify an otherwise complex story about Earth's geophysical past, since schools of thought are emerging that explain Ice Age genesis and ice age formation from different perspectives.

So far in this chapter we have been discussing El Nino-type events, and this has been done for a purpose; they exemplify more than anything else the concept of the chaotic runaway feedback. One could summarize this chapter so far as saying, in effect, 'Here we have something that starts off very small and turns into something uncontrollably big'; or instead, 'Here we have something that goes from west to east and suddenly turns round the other way'. They are a phenomenon described by mathematicians in the field of non-linear dynamics (an everyday example occurs in the sensor that responds to room temperature in a central heating device). Such real-world phenomena also seems to apply to the creation and dispersal of Earth's ice. 'First Earth has much ice, then it has little'.

This stark geophysical truth has been suspected for a long time, but it is somewhat disconcerting to have this truth rammed home so conclusively by science. The message for climate researchers, then, derived from some remarkable discoveries from sedimentary and ice-core evidence in recent years, is that perhaps Ice Ages, as well as being the product of cosmic events, (see Chapter Seven) can be internally created, and the Earth can switch back and forth between them and interglacials more or less instantaneously.

Recall that scientists, when speculating about past Ice Ages, still have much testable raw material at their disposal. The Earth's ice represents a unique historico-geophysico archive, and scientists have long been probing its mysteries. Ice hunks are not just records of living creatures and biomass from which we can make guesses as to the kind of climate that existed in olden times, to support them and it, but an actual record of the atmosphere itself. And, as we shall see, the rock sediment beneath the ice also tells its own story. Hitherto the problem was always one of trying to read the record.

The continual sub-zero temperatures of the poles naturally arise from the limited amounts of solar radiation they receive. But there is often enough moisture in the air to cause snowfalls, and thus build up fresh ice. As each layer of snow is compressed under the weight of further falls it becomes progressively denser. The air trapped between the grains of snow soon gets squeezed out, leaving minute air holes. The crystal structure starts to change, and stratified snow layers build up, ultimately to become true glacier ice. The extra dimension of air circulation needs to be taken into account in explaining the permanence of ice. The key factor is the cold, dry climates in the south polar regions that don't melt ice but don't make it accumulate either. Eugene Domack and his colleagues at Hamilton College, New York, suggest that katabatic winds howling across ice sheets play a major role in preserving snow. These winds arise when air cooled on high ground becomes dense enough to literally blow fallen snow away as the air flows rapidly downhill. As the world warms slightly, the winds abate, and more snow survives.[28]

Actually measuring recent snowfalls in Arctic or northern regions is virtually impossible because of snow drift. Snow and ice often stick to sensors, and modern techniques to thaw them out do not come cheaply. Some scientists have suggested making the sensors and outer structure of a system from hydrophobic materials, which both repel water and reduce the adhesive strength of ice.[29]

Other types of optics that try to take measurements of sunlight glimmering through snowflakes need considerable power to keep them free of ice[29]. Measuring the depth of snow can be misleading, too, as its density varies. Assessing its water content is not easy, either, because slivers of ice may reside inside the layers of snow. Paradoxically measuring some of the gases, acids and dust of olden times — in other words looking at some of the ingredients that we think today cause climatic change — is a rather easier task. These elements get trapped in the expanding ice caps when they are trans-evaporated into the atmosphere and returned in icemelt or rain.[31]

Other clues reflecting the growth and death of vegetation come from pollen spores found in the icecaps. They act as chemical fingerprints, preserved each year as the annual layers gradually drift downward in the ice. Electrical conductivity is also used as a test for atmospheric temperature, since it can indicate the level of dust in the atmosphere.[32] Dust is an indication of wind strength, CO^2 and methane concentrations, and also gives clues as to past temperatures.

Oxygen isotopes found in ice can reveal a great deal. The isotopes O^{18} and O^{16} are found in fossilized shells of marine animals such as forams (tiny unicellular creatures with shells of calcium carbonate). They have different masses, and how they distribute themselves in snow crystals depends on air temperature. During extended cold periods the lighter O^{16}, instead of readily evaporating from the seas and then returning in icemelt or rain, gets trapped in the planet's ice.

About 8,000 years ago the south polar oceans were more free of ice than they are now. Some sediment and ice cores showed salinity to be down 10 per cent on current levels, suggesting an iceberg melt. In warmer times the ratio of the two isotopes is less pronounced. One drawback is they do not reflect a good average temperature, since you can get heavy snowfall in warmer spring months.

The spores, dust and isotopes buried in the ice are extracted in long screws of solid ice several metres in diameter drawn out of the

Earth. One ice-drilling exercise dating back to 1963 measured some 100,000 years of changing temperature, and the Vostok core drilled into east Antarctica by Soviet engineers in the early 80s covered 160,000 years. Analysis of a 3,000-metre long ice core drilled from the summit of the Greenland ice cap recently resulted in some surprising scientific news that hit the headlines in 1993.

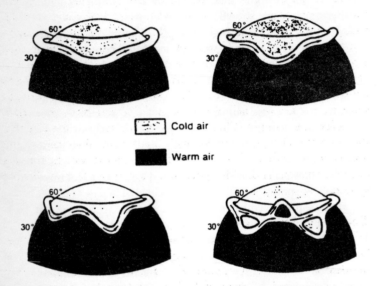

Prolonged cool or warm spells in the northern hemisphere are invariably due to the meandering jet streams which occasionally push down into lower latitudes. Stationary pockets of warm and cold air become pincered and dislodged, and held in place until they are cut off as the path of the jetstream straightens.

Source: Joe Eagleman, Severe and Unusual Weather, Van Nostrand Reinhold, 1983

The new Greenland Ice Core Project (GRIP) lasted from 1990 to 1992. It was undertaken by a European crew from eight countries, and took the record back nearly a quarter of a million years. It was the first project to span two ice ages and three interglacials.[33] And it was highly reliable, being drilled at a high point in the Greenland ice, where there is hardly any sideways slippage of the ice, with the very ancient ice layers held firmly at the base. The GRIP project was backed up by a team of American researchers undertaking what they called the Greenland Ice Sheet Project-II

(GISP-2).[34] Both scientific ventures each produced cores showing a fairly stable climate for the past several thousand years, and instability during cold periods.

But it was the speed of the change that was so remarkable. The discoveries that made the greatest scientific impact were the tests made at depths of between 9,100 feet and 9,400 feet , corresponding to the period between 115,000 to 135,000 years ago, when the climate changed some 10C or so both ways, cooler and warmer, taking only between ten and thirty years to do so, and remaining constant anywhere from 70 to 5,000 years.[35] There was, apparently, a cold snap in the 14th and 15th centuries that wiped out the 400-year old Norse settlements in Greenland. Other exceptionally cold periods in history, such as 1695, 1740 and 1816, and which brought subsistence crises to much of Europe, had fluctuations no more than 3C below the norm.[36] What was also unexpected was the fact that the Antarctic changes were so large and abrupt, since one would normally expect this only of the North Pole region, which, in terms of ice formation, is the more dynamic part of the Earth's climatic system.

The Speedy Ice Ages
What could have caused such rapid climatic reversals? The most bizarre theory decrees that a build-up of ice at the poles in prehistoric times may have altered the Earth's motion. This in turn may have changed the shape of the Earth and may have triggered the puzzling shifts in the frequency of ice ages.[37] This theory has not succeeded in overthrowing established orthodoxy but remains a serious threat to it.

In a seminal paper published in 1976 John Imbrie and his colleagues announced that the various ice cycles — the methodical stacking of snow on cold continents and its periodic melting — could be matched up to three cyclical fluctuations in Earth's orbit: the tilt of the planet's spin axis, the orientation (or 'wobble') of the axis, and the shape of the orbit itself. Together these cycles affect the amount of sunlight reaching Earth's surface at each latitude and season: the orbital tilt and wobbles act as 'the pacemaker of the ice ages'.

This theory was hardly new, being based on calculations made earlier this century by the Serbian mathematician and astronomer Milutin Milankovitch. Remember that periodicity in weather and climate must have some astronomic cause because all objects in space have gravitational and magnetic fields arising from their orbiting, spinning and rotating characteristics.

Milankovitch determined that the variations in Earth's orbit and axis of inclination happen at cycles of 23,000, 41,000 and 100,000 years. The axis tilt of the Earth is not constant, as its direction performs a full circle in space during the first, shorter cycle. In addition the Earth travels round the Sun, slowly shifting its perihelion point, and gradually nodding from being more upright to less upright, over the second cycle. It is this tilt, over 41,000 years, that it is the key, Milankovitch believed, to the Ice Ages. The more upright the tilt the more evenly distributed radiation the Earth will receive. This means cooler summers and milder winters in the north and south — ideal for building up the ice caps. Finally, over a period of 100,000 years the Earth's orbit around the sun becomes more elongated, taking the planet further away from the source of the sun's heat.

All the same, what happens to the northern hemisphere is more important than what happens in the south, since there is more land at high latitudes in the north to support vast ice sheets during a cool summer. 'It's always cold enough there in the winter to make ice', explains Imbrie, an oceanographer at Brown University. 'What counts is how much melts in the summer'.[38]

Taking off from Milankovitch, Imbrie and colleagues worked on an ice age theory that brings the three cycles together to conspire to keep the North cool for a long time during the summer; interglacial periods, like the one we're in now, happen when the northern summers get hot enough to melt the glaciers.

However, new theoreticians have now added an extra feedback; one that explains why the Earth should 'nod' in its orbit in the first place. Firstly, the Earth is not a perfect sphere: it is actually slightly fatter at the sides. But it was even squatter millions of years ago when the ice at the top and bottom of the planet was virtually minimal. This assymetry allowed the gravity of the moon and sun to gradually alter the axial tilt. However during an ice age, with the build-up of ice at the poles, the Earth becomes more spherical, so the cosmic influences lessen accordingly.

Bruce Bills of Nasa's Goddard Space Flight Centre in Greenbelt, Maryland, believes he is the first scientist to take this crucial repercussion into account in simulating earlier ice ages. In the past researchers have, of course, taken axial-tilt into account, but have assumed that it remained roughly the same in its terrestrial impact during the past million years.[39] But Bills found that this value changed when the asymmetrical features of the Earth were eliminated. A rounder Earth meant that over longer periods of millions of years the axial tilt cycle gradually increased.[40] Normally the tilt, which at present is 23.5 degrees, varies by about two degrees during the 41,000 cycle.

This may mean that Milankovitch's 100,000 year climatic cycle due to the eliptical orbit around the sun may be the most important cycle since the other two-degree oscillations would have had less of an impact once the ice caps became a permanent feature of the Earth. Hitherto it was thought that the 100,000 year cycle was the least important because the changing impact of solar radiation on Earth, merely as a result of a slight elongation of the orbit, was thought to be insufficient to result in climatic changes.

Yet even newer data add an intriguing prehistoric gloss to the ice cycle theory that not even John Imbrie and his colleagues took into account, and which at the same time have serious critical implications for the carbon dioxide-as-climate-controller argument. Both the creation and destruction of ice packs seem to have been the product of a warming, either in the atmosphere or in the oceans, in the case of the former, or from positional and geotectonic features in the latter.

First the ice build-up. During the Ordovician Period, some 440 million years ago, there seems to have been 16 times as much carbon dioxide in the air as now, and yet there was an ice mass near the South Pole that wasn't far short of the size of Antarctica today.[41] In the Cretaceous Period CO^2 levels were about eight times what they are today. Surprisingly Bruce Sellwood of Reading University suggests the climate was no warmer than it is now.

The reason for this was deduced from an analysis of sedimentary rocks from 95 million years ago. It seems that the exact location of the continents in prehistoric times was somewhat different from that prevailing now. The survival of a permanent ice sheet depends not so much on extra snow falling in winter but, as we have seen, what potential the summer warmth has for melting the winter snows. So, to get an ice age started you need generally flattened-out temperature gradients. Scientists have found recent evidence of glaciation in the south of Greenland long preceding that in the north. Hans Christian Larsen of the Danish Lithosphere Centre in Copenhagen believes this is because the south has mountains and heavy rain which make ice crystals grow faster than in the dry north.[42]

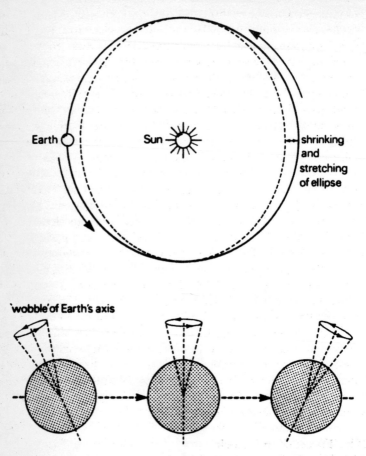

The Milankovitch cycles: The 'wobble' in the Earth's orbit takes place over a cycle of 23,000 years; the tilt in the Earth's axis relative to its plane of orbit takes a further 41,000 years. The Earth's orbit around the sun (above) completes a further cycle in 100,000 years.

Source: Antony Milne, *Our Drowning World*, Prism Press, 1989

The key to the Ordovician ice sheet was Gondwanaland; prior to continental drift all the continents were melded together roughly where the present south pole is. Then the Earth had extremes of continental climate in the interior. Along the southern edge of

Gondwana, close to the Antarctic ocean water, the weather was intensely cold. Although the interior was hot the southern margin of Gondwana accumulated three feet of snow each year, eventually to become part of the south polar ice cap. The glacier vanished only when Gondwana moved further north into warmer waters.[43]

The notion that a slight lessening of cold temperatures, over the glacial regions, might actually bring on another ice age was given a further boost by several mainstream scientists in 1992 when they found geological evidence that just before the last ice age global temperatures started to nudge up. This idea came from Gifford Miller, a prominent ice-age researcher at the University of Colorado.[44] Anne de Vernal of the University of Quebec at Montreal came to the same conclusion. So did Eugene Domack, whom we mentioned earlier, and a Japanese colleague from the Geological Survey of Japan. Using cores drilled by the Ocean Drilling Programme they found that the ice sheets expanded significantly between 3,000 and 7,000 years ago, when Earth was supposed to be recovering from its last ice age.[45] Anne de Vernal found that ice in the northern hemisphere began to spread south when the climate was at its mildest some 120,000 years ago, some thousands of years before the onset of the last Ice Age.

The salinity of oceans — an indicator of whether and how fast the water sinks to allow warmer, fresher, water to circulate — can also be worked out from the oxygen isotopes because the evaporation and transportation to icy latitudes is likely to yield O^{16} isotopes. More dilute rainwater is likely to yield O^{18}. Gifford Miller goes on to say: 'Everything we learn about past ice ages teaches us that the climate is more sensitive than first thought'.[46]

The Threat of an Icemelt

Now let us look at factors leading to an ice sheet break-up. The last ice age, which ended only about between 12,000 and 14,000 years ago, lasted for more than 60,000 years. But the ice age was not featureless or continuous, and there is a growing belief that the ice age was punctuated by many mini-interglacials and mini ice ages along the way.

John Andrews of Colorado University said an analysis of rock fragments suggested that they seemed to come from the centre of the ice sheet, hinting that the centre had collapsed and flowed outwards. Internal geophysical instabilities, rather than cosmic factors, seemed to be the prime mover.[47] Probably the rock beneath the ice sheet was warmed by geothermal energy. There would have been an enormous surge of the ice through the Hudson Straight into the North Atlantic. Large chunks of the west Antarctic ice sheet would have moved — in other words there would have been a break-up of Antarctica, a nightmare vision even now for many doomsday writers. This in turn had a convulsive repercussion throughout the entire Atlantic, eventually to precipitate yet again another ice age in the northern hemisphere.

The Ross Ice Shelf in the Antarctic has always held a fascination for scientists. This is an area of slab ice larger than France, and looking in the atlas like a wedge of ice that has been hammered inwards from the surrounding Ross Sea. Curiously, it seems to be both resting on land and floating on the water. The shelves in effect pin down the rest of the ice sheet, and prevent it from breaking up and drifting apart. As the ice shelves build up from fast-flowing streams of ice a critical point is reached where they begin to 'calve' or break away into icebergs.

The key factor seems to be local melting and lubrication, helping the layer of ice to move. But it is often aided by gravity and the inclination of both the ice packs and the ground beneath, which may sometimes slope upwards instead of down.[48] The Norwegian Polar Institute is accumulating data on iceberg breakups — some have been reported as big as Cyprus. But most are too small to be seen by satellites. Other kinds of ship monitoring of icebergs have not continued for long enough to make reliable predictions about Antarctica, and much of the prediction is now done by computer modelling. Bill Budd of the University of Tasmania has recently said that any global warming will take more than 100 years to have any effect on the South Pole ice. Whatever happens there seems to be a constant calving and replenishment of ice as snow continues to fall inland. Melting, he says, will not begin to overtake thickening for well over 500 years.[49]

A more catastrophic version of the ice break-up theory suggests that the weight of ice sheets might well have shattered the mantle of the Earth. Molten rock then oozed to the surface, beginning an eruptive volcanic cycle whereby carbon dioxide was pumped violently into the atmosphere from the deep vents leading from the hot bowels of the Earth below.[50] Shortly after, there would have been a renewed build-up as the ice sheet thinned and the sediment refroze, continuing a thawing and freezing cycle over a 7,000-year period.

Location factors arising as a product of continental drift over millions of years — can also result in other climate anomalies. We have seen that Gondwanaland may have melted its glaciers by drifting away from the South Pole. The Cretaceous temperatures, analyzed by Brian Sellwood and his colleague Paul Valdes, were little different from today's for precisely the same reason: the break-up of the land mass and the emergence of many new shallow seas. Nevertheless, the polar regions were a lot warmer than today. Valdes thinks that an embryonic Gulf Stream, the warm-water conveyor belt, may have been shallower, which would have more quickly transported heat out of the tropics. This is because the geographical distance, in the Cretaceous, between America and Europe was much smaller: 'There wasn't much of an Atlantic. So maybe the water didn't go down deep, it just recirculated and warmed up more and more, and so you didn't have such a big temperature gradient'.[51]

So an early Gulf Stream, like the present one, looms large in our explanation. Wallace Broecker, of the Lamont-Doherty Institute in New York, believes that warm ocean gyres may well be implicated in the formation of ice ages if the chaotic principle succeeds in magnifying things out of all proportion. The clue, says Broecker, is the North Atlantic and the amount of meltwater draining off into the warm Gulf of Mexico. He takes as read, as do many other scientists, some orbital factor responsible for getting ice ages started.

After that, however, the terrestrial cycles take over, jamming the Gulf Stream and cooling the north still further. When the climate inevitably reverses itself, new meltwater flows down to the Gulf, and the more northerly St Lawrence Seaway finally becomes ice-free. But the more frigid fresh water (which does not sink easily) just piles up and causes a temporary cold reversal in an already warming scenario.[52] This accounts, says Broecker, for the frequent mini ice ages that seem to occur for no orbital reason, such as that which occurred some 3,000 years ago.[53]

Broecker is a disciple of influential German scientist Hartmut Heinrich who has initiated a whole new train of research both in Britain and America, and brings us closer to understanding how SSTs can greatly influence climate. Study of micro-fossils dragged up in 1988 from seabeds in the north-east Atlantic by Heinrich's team at the German Hydrographic Institute proved there were several short periods during the ice age when rock fragments were dragged southwards via erratic iceberg movements. Bits of churned-up sedimentary rock tell us as much about ice age reversals as the GRIP cores, as ancient iceberg routes from the poles into the North Atlantic can be traced.[54]

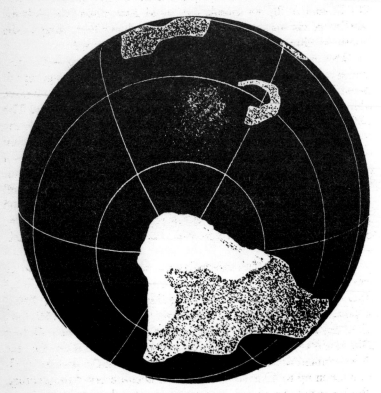

Continental Drift affects climate: glaciers survived at the South Pole until the Ordovician super-continent moved further north into warmer waters.

Source: Discover magazine US

Thousands of years ago silty sands and gravels were laid down, having been ploughed up from the Antarctic land mass by glaciers and ice sheets, and which now remain beneath fresh ice shelves. Icebergs in the Atlantic meant that salinity had dropped. As a result it affected the growth of phytoplankton, but at a much faster rate than once thought.

Some scientists are suggesting that perhaps similar climatic reversals — from warm to cold — are happening now. Broecker's theory, originally formulated in 1989, was recycled by himself and his co-scientist Claes Rooth, acting director of the Co-Operative Institute for Marine and Atmospheric Sciences in Miami, in 1994.[55] Rooth said a flip in circulation could slow down the Atlantic northward circulation, and this could bias the CO^2 warming in the south, or eliminate or reverse the warming in the north.

What is worrying is that this seems to be happening — the next ice age may arrive sooner than we expect. Several different groups of scientists seem to be saying this. Of crucial importance is the state of the world's existing stock of ice. Is it accumulating or melting? The difficulty is that an accumulation of snow, or stasis in regard to the present ice sheets, means little by itself, hence proving once again what a difficult science climatology can be given the importance of ice and SSTs to the subject. For example, Japanese research evidence that the sea level rise some 8,000 years ago, when the world was well into its post ice-age interglacial, came to an abrupt end about 2,000 years later may support Domack's observation that ice sheets were growing about 7,000 years ago, when the world was on average 2C warmer[56].

Finally, there is now a solid belief among glaciologists that the ice sheets of the world respond to long-term changes over tens of thousands of years. Climate warming alone, as implied by the IPCC report, doesn't seem to fit into the scenario either to explain this long time lag nor, surprisingly, much faster rates of ice melt. Robert Bindschadler of Nasa told the American Association for the Advancement of Science meeting in 1992 that in the past sea levels have risen up to 40 times as fast as has been observed this century. But it was not caused by ice melting so much as having something to do with the increased flows of solid ice: 'The dynamics of ice sheets can change very rapidly and I am increasingly concerned that the basis for believing that the ice sheets are stable may be invalid'.[57]

Perhaps there is no reason to think that the next full ice age is upon us, although a shorter episode of freezing conditions could happen at any time. The last interglacial was actually warmer than this one, and there is some evidence that it was more unstable. It is becoming clear from this chapter that the greenhouse effect, while warming up the Earth for a short while, may set in train events that could plunge the Earth into an imminent sudden chill.

FOOTNOTE REFERENCES:
Chapter Five: The Reversing Climate
1.   *Future Weather*, John Gribbin, Penguin, 1982, p72.
2.   New Scientist, Paul Simons, 7/11/92.
3.   Ibid.
4.   *Our Drowning World*, Antony Milne, Prism Press, 1989, p78.
5.   Future Weather, op cit, p31.
6.   *Searching for Certainty*, John L. Casti, Abacus, 1993, p79.
7.   New Scientist, 11/11/89.
8.   New Scientist, 2/2/91.
9.   Guardian, 11/8/94.
10.   New Scientist, Fred Pearce, 15/1/94.
11.   The Times, 2/5/91.
12.   New Scientist, op cit, 15/1/94.
13.   Ibid.
14.   Ibid.
15.   New Scientist, 23/7/94.
16.   Discover (US), May 1994.
17.   Searching for Certainty, op cit, p80.
18.   New Scientist, 4/2/95.
19.   New Scientist 2/2/91.
20.   Observer, 4/3/90.
21.   New Scientist, op cit, 7/11/92.
22.   The Times, 2/5/91.
23.   Sunday Correspondent, op cit, 28/1/90.
24.   Ibid.
25.   New Scientist, 23/4/94.
26.   Ibid.
27.   New Scientist, 8/8/92.
28.   New Scientist 23/11/91.
29.   Ibid.
30.   Times Magazine, 11/12/93.
31.   Sunday Times, Hilary Lawson, 12/8/90.
32.   New Scientist 28/8/93.
33.   Times Magazine, 28/8/93.
34.   Ibid.
35.   Ibid.
36.   New Scientist 23/4/94.
37.   Discover, (US), May 1989.
38.   New Scientist 23/4/94.

39. Geophysical Review Letters, vol 21, p177, April 1994.
40. Discover (S), December 1994.
41. New Scientist, 1/10/94.
42. Discover, op cit, December 1994. vol 345
43. Newsweek (US) 23/11/92.
44. New Scientist 8/8/92.
45. Newsweek, op cit, 23/11/92.
46. New Scientist, 4/9/93.
47. New Scientist, 25/9/93.
48. Ibid.
49. Newsweek, op cit.
50. Discover, op cit., December 1994.
51. Nature, vol 372, p82, November 1994.
52. See *The Fate of the Dinosaurs*, Antony Milne, Prism Press, 1991, p224.
53. New Scientist, 4/9/93.
54. Daily Telegraph, 11/6/90.
55. New Scientist, op cit, 8/8/92.
56. New Scientist, 28/8/93.

Chaper Six:

CAN CLIMATOLOGISTS PREDICT?

The 'climatic flip', the Greenland ice cores, and Bill Burrough's dispiriting pursuit of weather patterns, all point, as I have suggested, to *chaos*, both literal and theoretical. We remain what we have always been — creatures heading blindly into an unknown future, with even our computers peering into a chaotic abyss.

This is worrying, because climatology, one of the physical sciences, has taken upon itself the task not just of measuring changes in climate, but of also explaining why and how it will change in the future.

Understanding how the world 'works' involves developing a law to explain a set of *observations*: this is the *experimental* side of scientific endeavour. The other aspect of science entails the formulation of theories to explain a set of laws. This involves a jump to a higher level of explanation that takes in lower-level laws, such as Newton's theory of mechanics, which explained not only Kepler's laws of planetary motion but also included Galileo's findings about the motion of balls rolling down an incline.

The two procedures — experimental on the one side and the theoretical on the other — have an interplay back and forth. Hence the laws of nature that underpin theories are also used to explain empirical laws which are based on continuous observations of what happens in the real world. This can be illustrated thus:

Observations → empirical laws ⇄ laws of nature → theories
　　　Experiment　　　　　　　　　　theory [1]

Let us, as we are dealing with climatic issues, take the carbon dioxide argument and see how well it conforms to the scientific process. Scientists can observe that carbon dioxide has a warming potential because it can trap and deflect heat arising from the surface of the Earth. But this observation is derived from laboratory experiments with the carbon atom rather than from direct 'happenings' in the external world, like feeling warm on a sunny day. In this sense CO^2 warming is rather like reading off a setting on a voltmeter or pressure gauge. Thus, the sensory impression of warmth out of doors works against the carbon dioxide principle and in favour of an empirical law explaining why solar radiation warms the Earth.

Next, the theory of carbon dioxide warming can be formulated into empirical laws which decree that solid planets with elements of carbon in their atmospheres will be warmer than those without them. But it is when carbon dioxide is put into the second part of the equation, concerning laws of nature and theories, that it reveals inadequacies. Empirical laws do not automatically become laws of nature, since even regular clusters of observations can support events that *could* have occurred but do not always. It is fair to say that other aspects of nature have more of an inevitability about them — such as low atmospheric pressure producing wet and windy weather — than the CO^2 theory, largely because of the problem of feedbacks and multi-causal explanations. Furthermore there is the hazardous element of unpredictability that creeps in concerning long-term events, of which I shall have more to say later.

Finally, in relation to Newtonian-like theories, CO^2 fails completely since it remains with the status of a lower-level law that cannot successfully be incorporated into a higher set of laws which explain how planets with oceans and ice packs warm and cool themselves.

Climate predictions can be more or less right or more or less wrong. A forecasted warming could come in fits and starts, thus jeopardising early conclusions about the future of the climate. Scientists would simply have to wait it out, as if an experiment needed just a little more data that might just enable it to be formulated into a law of nature. Hence, until the experiment had been tried and tested one could not really predict anything.

So predicting future climates is unlike predicting conjunctions of the planets or phases of the moon — although as we shall see in the next chapter, many scientists think that perceived regularity might well be astronomical in origin. Weather-changes, on the other hand, occur within defined and known parameters which make prediction, in theory, easy. Even forecasting events such as stock market movements have a 'horizon of predictability' so near at hand as to give science no advantage over intuition.

The *statistical* nature of climate change poses more problems for scientists than they like to admit. Changes can only be inferred from statistics which take a long time to gather. Observations can be made with electronic or mechanical instruments and not just with the human eye. But this is one of the scientist's greatest problems: the unavoidable need to make everything abstract, invisible and data-driven.

Belief in the invisible is undoubtedly hard, and is in the same league as the scientific metaphysics of the Big Bang or black holes. 'Weather', says Mike Hulme of the CRU, 'can be seen but climate cannot'.[2] In other words we experience floods, storms and droughts directly, but no one ever lives long enough to personally 'experience' climatic change.

Worse, the faith of scientists in human technological prowess may be excessive. The longest satellite-time series relate merely to the constituent parts of the global climate and some of them have been around for less than 20 years. We would surely need at least 100 years of continuous monitoring using the same interrelated systems (rather than different instruments doing their own thing) before we could tell for sure how the climate was behaving.

Here ground-based instruments score over satellites — simply because they have been around longer. The 'eye in the sky' has its limitations, and sometimes instruments fail, but there is often lack of continuity or comparability. After the enquiries into the British 1987 storm Peter Swinnerton-Dyer of the University of Cambridge and Robert Pearce from the University of Reading Meteorology Department said that when it was decided that meteorologists were only partly to blame[3] for failing to predict it. The truth was slowly dawning that literally infinite amounts of information about the basic components of the weather, especially the hydrological aspects, at every part of the globe's surface, would be needed to avoid such failures.

Statistics are part of maths, and maths, as we shall see, are used in models, and models are used to predict future trends. But the models rest on earlier scientific precepts — that is, they are derived from scientific rules and laws which are invoked to explain why things are as they are: why, for instance, the Equator is hotter than the North Pole. Prediction and explanation are the main goals of all scientific endeavour: it is a reality-generation game. Yet greenhouse predictions are bedevilled by complex inputs relating to rainfall, cloudiness and ocean currents which could greatly affect snowfall at high latitudes which in turn could either cancel out or reinforce the greenhouse effect. T. P. Barnett of Scripps, refers to 'time-dependent' biases in the measurement technique, such as cloud cover, volcanic eruptions, and sudden ocean warming: 'Each item could raise or lower the estimated global temperature. But if we see a change in the temperature, how do we know which process to attribute to it?'[4]

Statistics and prediction come together in a not particularly sound relationship in short-range forecasts. Forecasters, even without consulting their charts, say they will expect a continuation of mild, settled weather because the odds are highly likely that this will occur. An error in a three-day forecast is only 20 per cent less than the *persistence error*, which is the error made by predicting that tomorrow's weather will be the same as today's. Zero-change predictions are often used as a marker against which to check the validity of science-based forecasts. Errors in forecasting for a few days ahead have been falling consistently over the last decade or so, probably being 'out' by only 20 per cent or so, and yet the 'persistence error' can rise to 75 per cent by forecast-day six. Generally, though, there is a limitation to the length of time meteorologists can make accurate predictions about the weather.

Prediction sometimes rests on *causation*: things happen because of clearly defined antecedent factors. Unfortunately not all laws are causal, and might not actually be required to be so. The apparent insistence on cause-and-effect reasoning might be a subconscious tendency of humans to recreate a meaningful world where causes always precede effects, even if they are accidental or unexpected. This is why the paranormal can be so psychically disturbing: things may happen with no known logical antecedents (and are part of private and hence untestable experiences), and if they were to occur continuously they would make the universe a meaningless and terrifying place.

Causation, of course, must not be confused with correlation and explanation. Indeed statistical correlation in regard to changing weather patterns are fraught with hazards, since one could argue that weather and other events do go hand in hand. For example, people in the west wear shorts and skimpier clothes in seasons of warm weather. But other correlations can be highly misleading. A scientist writing in *Nature* magazine once warned about the statistical hazards of perceiving correlations and deriving the wrong conclusions from them.[5] There was apparently a 'positive relationship between CO^2 levels and rising temperatures'. But *Nature* went on to add: 'The population trend of San Diego is also positive over the same period (1958 to the present) but no one is suggesting that it is responsible for the apparent increase in global temperatures'.[6]

Chaos and unpredictability are often associated with randomness; with events revealing no clear pattern. Yet randomness has no objective meaning and is not a mathematical problem. Empirical laws exist to help organize seemingly random observations into some meaningful pattern. This is where periodicity in climate is more bothersome — since rhythmic peaks in weather events might well be correlated to sunspot activity which peaks in a cycle of about 11 years (see the following chapter) but might well conflict with other empirical laws where periodicity is excluded from the overall theory explaining a set of laws.

In fact if you do try to eliminate all the known periodicities, and there are not all that many, a lot of variability still remains in the weather system. Indeed, with regularity you could actually dispense with weather forecasts, and there would be no new weather extremes either. To some extent this happens already: the seasons and months within the seasons have temperatures and rainfall that fall within an average range of data and make it easier to build in zero-change assumptions.

Early Measuring Hardware

On the whole, it should be stressed that the success of meteorology is very high, surprisingly so, because most of the weather forecasting instruments in use today are still the simple types, or modelled on the simple types, used in the first half of the century to meet the demands of shipping and aviation.[7]

At a typical site, air temperature can be read from an ordinary domestic mercury thermometer surrounded by screens of slatted wood which shield it from sunlight. This type of screen was first designed by Thomas Stevenson, the father of author Robert Louis Stevenson, in 1866. Air humidity is calculated behind the screens by an ages-old technique — transevaporation, a principle underlying simple domestic cooling devices, where a fan blows across a moist grid of porous paper strips. An instrument with a wick dipped in water keeps the instrument cool, and the comparative drop in temperature is then used to measure air humidity.

Wind direction is still checked by looking at streamlined versions of a church weather vane. Wind speed is measured by a cup anemometer, an object with three cups rotating on a shaft, invented in 1846 by a clergyman.[8] Sunshine hours are recorded by an 1880-type Campbell-Stokes instrument, where a glass ball concentrates the sun's rays into a ray of intense heat and literally burns a line into a strip of calibrated cardboard. The simplest device of all is a calibrated phial for collecting rainfall, although in recent years the search for a better rain-gauge, with more aerodynamic funnels and so on, has continued. Even so, most rain-gauges have little protection from the wind. Indeed as rain never falls evenly even into two rain-gauges placed only a few metres apart, there is no such thing as a perfect rain collector.[9]

One factor which betrays an inevitable lack of rigour is the bias towards land-based instruments. Weather stations take some 60,000 measurements a day, but the type of site chosen is not evenly distributed over the globe. More sites are to be found in populated areas, and in the northern hemisphere, while the two-thirds of Earth's surface covered with seas is virtually unrepresented.

Most weather stations are found in or near cities. Robert Balling, a climatologist at Arizona State University, has found an urban warming bias in excess of 0.5C. This is a conservative estimate: Paris is now about 2C warmer than its suburbs;[10] even villages of a mere 300 people can produce localised urban warming of 0.3C — almost as much as was supposed to have been occurring in the last century.

The reason for this phenomenon can be explained with the aid of some simple physics. Different surface materials have varying levels of heat conductivity. Spongy, frequently moist, arable soils generally have a low rate of heat conductivity, and a higher rate of light reflectivity. In addition the superstructure of city buildings and pavements, the tarmacadamed roads and concrete walls, act as a giant thermostat, soaking up the sunlight and retaining the heat of the day.[11]

Ideally measurements should be taken in a well ventilated and specially designed shelter over open mown grass. Failure to do this in the past has cast suspicion over many famous measurements that have gone down in the British weather record books, such as the 100.5F (38.5C) observed at Tunbridge Wells on 22 July, 1868 when the measuring instrument was more susceptible to direct heating from the sun.

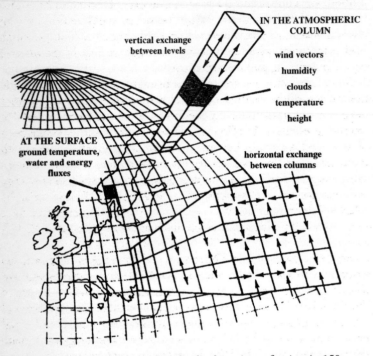

A 3-D model of the global atmosphere divides the entire surface into its 150 square km units, and upwards to the stratosphere into 15 vertical segments. Computers then work on the numbers gleaned from the grid points. The acid test of any model is how well it can mimic reality.

Source: John L. Casti, Searching for Certainty, Abacus, 1993

However the urban bias doesn't apply universally: for instance data from the former Soviet Union showed urban stations being cooler than surrounding rural ones from 1953 to 1967.[12] Mainland China showed a similar cooling in the 1960s.

Old fashioned ocean temperature measurements are flawed in similar ways. Many oceanic readings come from islands, by definition different from their watery surroundings. Instruments are often situated near airstrips, to provide relevant information for pilots. The towering height of postwar ship decks was surprisingly found to have an affect on the accuracy of measurements, as air temperature falls with height and produces a spurious cooling. Some ships' thermometers were even suspected of being exposed to direct sunlight or were rested on warm structures. Disparities have also been caused by operatives taking the ocean temperature, even in postwar years, in canvas buckets dipped over the sides of ships, which inadvertently allowing the water to cool appreciably in the interval between sampling and measurement.[13]

Since 1987 microelectronics has brought precise and automatic monitoring of the weather ever closer. Raindrop-detecting weather radar is now commonplace, as we can tell from the comments of TV weather presenters. In effect, the radar confirms earlier predictions of wet weather made by other instruments. Electric vanes with miniature platinum resistors are slowly replacing the older type of vane, and new Earth-observation satellites can measure solar energy and distinguish it from ground radiation[14], which can then be relayed via out-stations. Data recorded on one lithium battery for a year in a small memory can be obtained at a cost of less than £1,000 a year.

Bob Charlson of the University of Washington in Seattle is working with several colleagues at the NOAA laboratories who are mounting a shipborne expedition to get a more complete picture of the boundary between a sulphate-laden northern hemisphere and the more pristine south.[15] There is, however, no coordinated programme among the rich nations who are beginning to automate their data networks. The World Meteorological Organization, based in Geneva, reacts conservatively and cautiously to mooted changes in long -established practices, partly to keep costs down and partly to avoid creating too much chaos while a change-over is implemented.

Early Computerization of Techniques

Weather forecasting is a high-tech science. Britain leads the world with this science, as befits our national fixation with the weather, not only in the quality of its meteorological knowledge but also in its applied provision of meteorological services to important clients like the government, airline corporations, maritime and military organizations. Privatization in 1990 meant a high dependence on commercial contracts, like the one with the Ministry of Defence. It also meant that the Met Office is now under some pressure to increase the commercial sales of its 'products'. Future services the Office might sell will reflect the fact that it is now an environmental agency as well as a meteorological one.

Julian Hunt took over as Chief Executive of the Met Office from Sir John Houghton in 1992, and was, like him, an outsider who was not even a meteorologist, being a former professor of fluid dynamics at Cambridge. But Hunt's earlier career put him in touch with pollution researchers at America's NCAR, and his other work with a small consultancy company gave him added management skills useful for running the Met Office's 2,400-staff office.[16]

World weather news-gathering is a mammoth undertaking, and all countries co-operate in this through the WMO, which agrees international standards.[17] The Met Office in Bracknell is one of two World Area Forecasting Centres, and vies with the one in Washington. It gleans raw data from thousands of points around the globe. The complex equations loom ever larger: to predict London's weather you need data from both the Arctic and North Africa. To make a four-day forecast you need global facts from all over the surface to right up into the stratosphere. Hence the WMO becomes indispensable, and so does its vast array of computerized equipment.

Computerized meteorology has had two illustrious and noteworthy beginnings, the first starting in 1910, and the second 40 years later. Modern meteorologists are the inheritors of the tradition of Laplace, the 18th-century French mathematician who wrote of the possibility of an Intelligence that could understand all the constituent parts of Nature and all the forces impinging on them at every moment, and could hence make *total* predictions about the future state of every particle, object and human being. If we could do that we could develop a Grand Theory of the Universe rather like the cosmologists are trying to do today with their Theories of Everything (TOEs).

The Laplacean tradition was adopted in 1922 when Lewis Richardson believed we might be able to turn the Earth's blanket of air into pure numbers. Richardson was a physicist and mathematician, and during the First World War, while he was attached to an ambulance convoy, wrote a pioneering book called *Weather Prediction by Numerical Processes*.[18] Taking a solitary forecast in 1910 he drew up a checkerboard grid over Britain and north-west Europe, and assigned variables such as temperature, water vapour and wind velocity to each square on the grid, which also had a vertical dimension. However, his attempts to calculate how these variables would change over time ended in dismal failure, since the end-effects were a hundred times too large, resulting in unearthly super-hurricanes with wind speeds of 200 mph! Yet he was the first to suggest that the weather might well be successfully computable in the future.

In fact the computerized approach to weather forecasting is what the Met Office calls NWP (Numerical Weather Prediction). The mathematical, three-dimensional grid is a successful reality. No such machine could be built in Richardson's lifetime but, in recognition of his brilliant foresight and commitment to the idea, the wing housing the Central Forecasting Office at Bracknell bears his name.

The flaws in Richardson's calculations were really of a technical rather than a conceptual nature. His calculations could only, in those days, be done by hand, which meant that the checkerboard grids had to be large, since making them smaller would have increased, in geometrical proportion, the need for ever-higher levels of computerization and would have made the task too much for one man. Furthermore the observational data available in 1910 was obviously very primitive. And the time-steps in Richardson's model were longer than the time it takes for a sound wave impulse to move from one grid point to another. This brought about a computational instability which was a condition only discovered by scientists in 1928. Nevertheless, both meteorology and climatology now rely heavily on mathematical models which try to mirror the world abstractly, using the rules of Aristotelian logical inference generating new mathematical relations, on the basis that if A is true, and B is an instance of A, then B is also true.

The eminent Princeton mathematician John von Neumann realized, in the early 1950s, that the computing machines he was building were indeed ideal for calculating the motion of the atmosphere based on mathematical relations. He and some of the best brains in meteorology had the early computer models 'up and running' by the end of the decade. Neumann's brilliant team of young scientists was headed by Jule G Charney, who later turned the art of numerical weather prediction into a science.[19] Carney and von Neumann together gave the first computerized weather forecast in 1950. Later, in 1956, one of the Princeton group, Norman Phillips, made the first attempt to model the global atmosphere. Seven years on the NOAA established the Geophysical Fluid Dynamics Laboratory at Princeton. The fastest and largest computers then known to science were devoted exclusively to mathematical modelling of the atmosphere. The numbers in the computer models represent inputs such as solar energy, gases and physical feedback processes that 'take out' CO_2. The laboratory is staffed by a hard-working multinational team, including a young Japanese scientist and famous climatologist, Syukuro Manabe.

Computer Climate Models

The first true *climate* model was developed by Manabe and Richard T. Wetherald in the 1960s. Hence the die had been cast with the Manabe/Wetherald computer which has had significant conceptual consequences.

It is worth bearing in mind that a scientific model is an abstraction of some segment of reality — a model of a car or a plane, showing just the bare functional necessities. For example in meteorology the inputs, such as solar radiation, the state of the atmosphere, and the various albedoes and feedbacks, are converted mathematically into new quantities that conform to the laws of physics and chemistry. These then produce *outputs* — the weather forecast. One question bothering climatologists is how well this conversion process can be satisfactorily visualized. If belief in the invisible is hard, then belief in invisible changes is even harder, like trying to prove that eventually all black holes become white gushers or red giant stars turn into white dwarfs when no one can really prove that such astronomical objects exist.

If indeed climatologists are today all Smagorinskians, as John L. Casti, a mathematician from the University of California, says, the conceptual problems might be of a different order. Smagorinksy schools (named after Joseph Smagorinsky, a distinguished Princeton climatologist) are heavily enrolled with computer buffs who adapt weather-forecasting GCMs (General Circulation Models) and NWPs for use in climatology. However, there is a risk that climatologists, operating through a chain of scientific disciplines, lose out on their specialism. Many of them are oceanographers, but oceanography itself is not a single science but a composite of many basic sciences applied to the marine environment — biology, geology, chemistry, physics and mathematics.

Furthermore, the study of weather is a science that deals

essentially with phenomena characterised by *aperiodicity*, whereas the climate is *cyclic*. 'The difference', says Casti, 'lies in the fact that the two systems are qualitatively similar but far from quantitatively identical'. The 'spatiotemporal' scales of the two — weather and climate — are too dissimilar, and in effect are two different systems. Climatology, continues Casti, deploys both empirical and explanatory categories of models.[20] The first involves examining past trends to build up statistical explanations, and the second uses physical laws for prediction. These laws have mathematical equations, arising from Newton's second law, which then transform the inputs into the laws of thermodynamics and hydrodynamics. These in turn are used as outputs which are the variables the scientists want to forecast.

But this figures only in the short term. Weather prediction deals with phenomena taking place over hours, in regions whose spatial scale is measured in kilometres. Climate, on the other hand, involves a timescale of years, and grids encompassing regions the size of large countries *or* large oceans. Modern variants of climate GCMs can now only be created by large, expensive and scarce computers such as the CYBER-205 in the Central Forecasting Office.

The Cyber, solving about 1.5m equations a minute, was first bought by the Met Office in 1981. The later Cray YMP 8/32, costing £5 million, is four times faster than the Cyber, nearly doubling the computing speed[21], and has greatly dispensed with the need for staff, just as the use of computers everywhere in the industrialized world has drastically reduced the number of office workers. Computers can now produce a global forecast in just four minutes. Each forecast is for 15 minutes ahead, and builds on the last.

These computers can store limitless amounts of information in regard to both past and present weather dynamics. Altogether around 12,000 haphazard bits of data arrive at the Met Office every hour from diverse sources. Some 7,000 weather stations around the world take 12-hourly readings of atmospheric pressure, rainfall and wind temperature, which are then piped through the Global Trunk System to the Met Office. About 1,000 observations, twice a day, mainly land-based, are relayed from weather balloons as they drift out to space and relay their messages to Bracknell.

The number of calculations involved in working out the climate is truly enormous, and computers can take a few hours to do what humans would accomplish working with pen and paper some 15 years to do. Only a few research establishments can afford such mammoth machines. Famous climate institutes such as that at the Met Office, the CRU at East Anglia, the Goddard Institute for Space Studies and the NCAR are the principal ones.

The world, according to a computer model, is little more than a Richardson 3-D grid pattern. It bears the same reality to real life as does a computer game. A 3-D model of the atmosphere divides the entire surface into its 150 square km units, and upwards to the stratosphere into 15 vertical segments. This effectively reduces the atmosphere into 660,000 cubes. The boxes of the grid represent real Earth space of several hundred kilometres in width and several kilometres in height. There is a coarse mesh model of the global atmosphere, plus a fine mesh model of the North Atlantic area.

The more detailed grid, consisting of 4,000 points, covers, of course, a smaller area, and is repeated at 15 levels upwards to a total of 60,000 points at 75 km gaps. To keep the calculations within manageable limits the predictions refer only to where the lines of the grid intersect. It is the global model that is used for five-plus day forecasts for naturally it has a longer 'view'. For very short term forecasts artificial intelligence (AI) is sometimes used.

The computers then put all the numbers gleaned from its grid points through a series of differential equations based on a few elementary laws of physics. Real accuracy has now improved from two days ahead to four days in the past ten years, thanks to better number crunching and better observations. But after five days the probability of accuracy declines to below 50 per cent. Possibly in another five years or so we may have improved on this longer-range accuracy.[22]

In 1990 the Cray was known as the 'unified' model because it took in the greenhouse effect as well, whereas earlier models simply predicted the climate in linear fashion. Perhaps this is where the Met men have gone wrong. Even so the fastest computers would take several hours to come up with a simulated climate lasting just one year. The computer model for global forecasts has now been merged with the model for global climatic change, drawing all the sophistication of climate change into the forecast.

For yet greater accuracy one can imagine the need for many more multiples of ever more powerful super-supercomputers for every additional day and week of the forecast, costing many more billions of pounds. At the moment, says Nick Flemming of the Institution of Oceanographic Sciences Survey, he and his colleagues aim to be 85 per cent accurate on a 24-hour domestic forecast, but a 10-day forecast just as accurate is in the offing once a better observing system in place.[23] And by 1995 scientists are predicting accurate weather forecasts up to 12 months ahead.[24]

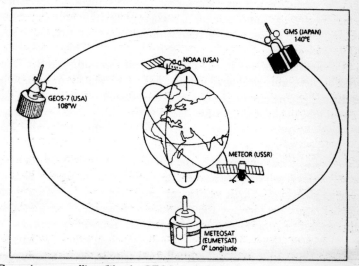

Geostationary satellites (like the GEOS-7) sit high above a fixed point on Earth. Polar satellites (like the Meteostat) have better resolutions and are always moving relative to the Earth's surface.

Source: Geographical magazine

In the meantime humans are not totally excluded from the forecasting exercises. Weathermen are still needed to interpret computer forecasts and apply local topographical knowledge, or just for spotting incorrect data.

Data may be distorted or inaccurate for many reasons. Information coming from remote stations or from local unpaid volunteers is the most suspect. It can all be a rather imprecise business because changes in any part of the grid can affect the entire weather system, and the unpredictable can always happen. Faults can arise from centrally located land measurements and poor data can originate from the patchy network of sea-based reports from oil rigs, ships and commercial aircraft. Finally, there will always be

limitations on what computers can do. The 'spin', the adjustments
and compensations, do take place.

There is one other major problem concerning the lives of
millions of poor people today: that is, the ability to forecast storms.
Do the new computers improve on this? Mike Hulme of the CRU,
when reviewing Bill Burrough's 1991 book *Watching the World's
Weather*, made the interesting point that improvement in forecasting
is not the same as improvement in accuracy, and human lives are not
necessarily saved (the latter requiring improved institutional and
political responses).[25]

The acid test of any model is how well it can mimic reality. The
present meteorological environment has to be related meaningfully
to the past environment. This is an exceptionally tall order. Even the
known industrial emission history of the past 200 years is not always
digested by computers, and most GCMs cannot properly take
account of the vital role that vegetation plays in climate modelling.
Plants can add water vapour to the air, and the reflective nature (the
albedo) of the surface varies when vegetation blooms or dies back.

One drawback is that operatives can't seem to make the grids
smaller, and thus increase the accuracy of the predictions, because
the computing power, even working flat out, is never enough. But it
is the small-scale detail from which the large-scale climate is derived.
Atmospheric turbulence and rainfall, for example, occur mostly on a
scale of kilometres, not hundreds of kilometres.[26] Sometimes factors
such as humidity and temperature are used as a marker for other
inputs, such as cloud formation, where the latter cannot be
adequately programmed for.

After the catastrophic English 1987 storm the Met Office report
blamed the duty forecasters on the night before for following the
guidance of their computer models too closely. These models were
too prone to error in a developing storm situation without adequate
upper and lower atmosphere measurements. Understandably
aircraft and ships would avoid such areas. Hence the grids became
less accurate. In fact two Met Office forecasts were in conflict with
each other, although this uncertainty wasn't fed into television
forecasts.[27]

Indeed, simply tracking the history of cyclones, which are known
to correlate with atmospheric pressure patterns, could yield equally
good results. One calculation of this sort gave a 1.8C warming in the
20th century, and compared well with the expected future warming

of some global models. Remember also that Robert Lindzen's main complaint about computers was that they couldn't accurately take into account latent heat transfer (see Chapter One). This aspect of climate study, based loosely on the theory that hot air rises and displaces cool air, possesses insufficient quantifiable precision to provide the necessary rigour upon which a whole raft of other calculations and assumptions must also be made.

The daily variation is a factor often overlooked. There is some evidence that maximum and minimum daily values have flattened out more since 1950. If the warming takes place during the night, as some have suggested, then evaporation will be less of a factor and droughts will become less frequent. As the growth of plants is inhibited by cold nights, the growing season will be longer with the onset of milder nights. Another problem arises because the models calculate everything from the median, from the equilibrium, forgetting that Earth may take hundreds of years to reach this ideal, and they may start to evolve their own large-scale laws which might no longer reflect the true situation.[28]

Some areas of science are simply too young or complex for modelling. There is always the tendency to simplify. But as Einstein is reputed to have warned: 'Make it as simple as possible, but no simpler'. One over-simplification is to identify a trend, assume it will continue, and then describe it by a very simple mathematical function such as a bell curve or an S-shaped curve. Precisely because they are so oversimplified these trends get the most media attention. For example Princeton economist Uwe Reinhardt once produced what he called 'the mother of all health-care forecasts' by extrapolating into the future in a linear fashion, so that, by the middle of the next century, it appeared that 50 per cent of GDP would go on health care. The biologist Paul Ehrlich's notoriously over-emphasized message about the 'population bomb' was also the product of simple linear forecasting, although most systems are too complicated for that.

These errors are partly due to over-reliance on 'curve-fitting'. This is a process using maths to describe a given set of data to within a given margin of error. A scientists will cite a number of examples of global warming, 'plot' them on a graph, and an equation describing the plotted points is generated on a computer. The equations are then used for predictions. But Douglas S. Riggs, a mathematician, reminds us that any set of data has more than one description, so that a number of similar curves, defined by different mathematical functions, can describe anything equally well, but each will yield a different extrapolation.[29] Each curve will level off at a different rate.

Unsophisticated use of maths curves, however, has probably done more harm to anti-Aids campaigners than to the greenhouse-mongers, since in 1986 well-qualified mainstream scientists were forecasting horrendous rates of HIV infection across the western world that never happened. The reputation of scientists is invariably sullied when this kind of thing happens, especially when they allow their conclusions to be rather recklessly publicised by non-scientists.

But unfulfilled predictions of that order are also due to other worrying trends in science: the incursion of politics. The Aids curve-fitting techniques had to assume at the outset that everyone in the population was at equal risk of acquiring HIV and Aids, instead of the infection remaining, as it still does, within very limited high-risk groups.

In the greenhouse predictions, similar curves also assume that measured increases in CO_2 have precisely defined climatic outcomes. This is often recognized, and compensatory adjustments are sometimes made to GCMs. Semi-automatic data filters can help here, such as the one recently set up by the Forecast Research branch of the Met Office. Individual observations can be tested against what would be expected from the general pattern in the area.

However, fine-tuning can hardly be said to add to scientific rigour. Climatologists Mike Jones and Tom Wigley once mentioned fine-tuning casually in passing, using the words 'systematically correcting the data both from land-based and from marine observations to eliminate potential sources of bias'.[30] They resorted to techniques such as comparing marine records made near land, and land-based observations made on islands and coasts, where one assumes more moist temperature conditions would equally apply. Remaining awkward differences were smoothed out, or eliminated with the use of jargon. For instance, land-marine differences were attributed to 'marine-measurement inhomogeneities', and mention was made of 'deriving correction factors by averaging these differences over many regions',[31] and that 'any remaining uncertainties must be blamed on poor global coverage'.[32]

Adjustments were once made to account for the heat island effect by Tom Karl of the US National Climate Data Centre, where he and his colleagues have set up the Historical Climate Network (HCN).[33] But here again, even if one accepts that the North American record cannot apply to the whole world, the HCN data shows an erratic cooling for most of the past 50 years.

Sadly, we must conclude that none of the climate models are reassuring. Stephen Schneider of the NCAR has even gone so far as to suggest that perhaps all the combined errors in climate forecasting might cancel each other out.[34] He said it is likely that the models were off by a factor of three, which would not negate altogether a marked temperature increase. Some would argue, however, that such a bias would totally invalidate the computer models.

Surely no engineer would work to the large grid magnitudes of GCMs if he had to base policy on them, because policy effects and outcomes would be that much greater than the modelled details. Yet the fact is that we are moving toward binding international policy based on conclusions drawn up by policymakers who often fail to distinguish between the levels of confidence they should place in what they want and what they have to work from. 'A system is not valid just because it gives you the answers you want', said former White House Chief of Staff John Sununu.[35] In fact, delegates to one of the final working groups of the IPCC in June 1990, in Moscow, complained that discrepancies created uncertainties in predictive power that led to a 'watering down' of some possible impacts.

A Chaotic Explanation
The practical problem for the greenhouse-mongers is whether minor inputs of warming gases could be the 'minor perturbations' that are likely to produce climatic disturbances. Computer models continue to programme for extra CO_2 inputs. Yet Tim Palmer, head of the Predictability and Diagnostics Section of the European' Centre for Medium-Range Weather Forecasts in Reading, Berks, cautions about doing just this.[36]

Furthermore the computer grids can't be improved nor can new physics be introduced because mathematicians believe we may now be at the limits of the predictive power available to a numerical system, and can do nothing to avert chaos. Chaos theory is worrying because it says, in so many words, that minor perturbations can result in giant catastrophic upheavals. But it is just as likely to be the outcome of El Nino events as CO_2 inputs.

Of course many leading meteorologists, like Sir Colin Flood, Director of Forecasting at the Met Office, say they knew about the implications of chaos theory for years before it became known as such in the mid 1980s.[37] Ian Stewart, a mathematician at Warwick University, says the Met men should have published their ideas before Robert Gleich, the science writer, did to much popular acclaim and financial advantage[38]. Few areas of science are as precise as classical physics and astronomy. Yet scientists knew they could solve the equations describing how two planetary bodies gravitationally interact, but not for three or more others interacting. They also couldn't use quantum mechanics, a highly quantifiable discipline, for describing photosynthesis. The limits of computer validity are hence strictly specified. Indeed some problems cannot even be written in computer argot. Roger Penrose, an Oxford mathematician, has suggested that the world is deterministic and non-computable. In short, climate, as well as human-related subjects such as economics, represents phenomena that are still unravelling in their own chaotic manner.

But why does chaos occur? The clue lies in what I have said about gravitational forces. Prior to the use of the term 'chaos', scientists used 'turbulence'. The problem of the onset of turbulence in fluid flow is one of the great mysteries of physics. For climatologists it was an irritating and worrying mystery for obvious reasons: the main method of analysis for meteorologists is *fluid dynamics*, because the atmosphere behaves like a turbulent fluid. Clearly fluids are not solids and hence cannot be expected to behave like solids. But the physicist trying to cope with the solid world still has his problems. Dynamical laws of motion applying to weights, measures, pressures, temperature and velocities are described by Newtonian classical physics. But Newtonian physics cannot help us make predictions about the future. The most important thing we can know about dynamical behaviour is where a trajectory of a moving object ends up. But this is something we can never know.

A *vector field* is like a system that entails a certain amount of meandering and circumlocution until events in the system reach a certain cut-off phase called a fixed point, beyond which no more meandering can take place. One can imagine a fly wandering over a desk in rough circles until it walks onto a flypaper trap placed at random on the desk. With a bit of luck the fly may meander forever without stepping onto the flytrap, repeating its wanderings in what is known as a *limit cycle*.[39] Catastrophes and population crashes — all fixed-point events — are part of this syndrome. The Gaia principle of biospheric self-regulation can be included in a limit cycle which avoids the catastrophic fixed point. The DMS situation we talked about in Chapter 4 suggests that when atmospheric temperatures rise, plankton start to thrive in the seas, then take in more CO^2 from the air. This tends to lower temperature, so that plankton populations begin to die back and soon an equilibrium limit cycle is reached. In other words, all trajectories tend to be *attracted* to this limit cycle, otherwise nothing would exist on Earth without coming to an abrupt and premature end. A further example would be the ratio of prey to predators in the wild, the rise and fall of either prey or predator having an inverse ratio to the fortunes of the other until an equilibrium is achieved.

The limiting factor is hence known as an *attractor*: the shockingly beautiful patterns often shown on TV science programmes, which large numerical systems are brought in to explain. The *strange* attractor traps trajectories, and has recently become better known because of the widespread use of computers. It is also part of chaos theory, because it forges a direct link between the idea of the complete Newtonian predictability of physical phenomena, and the unpredictability of randomness.

A dynamical system will not, having been 'wound up', just carry on forever with mathematical certainty. Even the planets whose orbits, can be worked out to the day, hour and minute, are still subject to the perturbations of other planets, as we have seen. A dynamic system can be thrown off course by a change either in the starting point or in the vector field. In the case of the latter it is like a human moving the flypaper into the safe limit cycle of the fly, thus trapping it for the first time in its life.

Fluids are not really dynamical systems at all, and are even more likely to be thrown off course. Fluid flow is characterised by the way a stream quickens into a torrent, or the way a dripping tap starts to gain on itself until the regularity changes, quite abruptly, into irregular spurts of water. A quarter of a century ago turbulence was thought to be due to an ever-increasing succession of higher and higher orders superimposed on periodic orbits. This was defined by something known as the Reynolds Number, and it would increase in line with flow velocity and viscosity. Soon a critical value would be reached at which smooth flow lines give way to small whorls, the

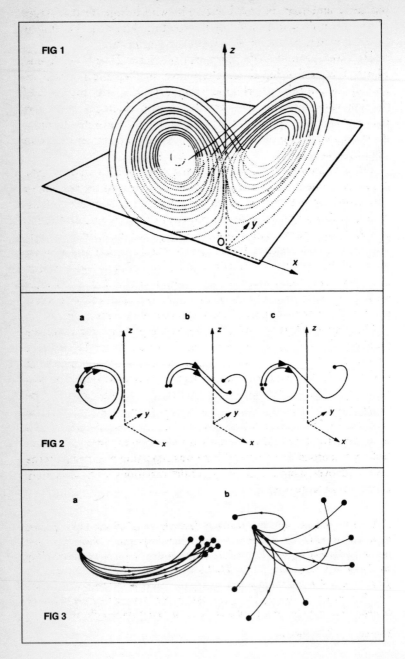

FIG 1

FIG 2

FIG 3

physical counterpart of limit cycles. The whorls beget further stages of whorls within whorls, with the flow pattern (known as the laminar flow) ultimately becoming all unpatterned turbulence.

However, around 1970, European physicists David Ruelle and Floris Takens developed an alternative mathematical model to remedy the unlikely 'whorl upon whorl' scenario. After the onset of the first whorl the cycle gives way to a quasi-periodic motion, a sort of mix of two limit cycles. At the next phase of the Reynolds Number the system enters a type of attractor, something hitherto unknown to classic physics.

Edward Lorenz, an innovative meteorologist at MIT in the 1960s, did an experiment nearly ten years earlier and brought the concept of 'attractors', something scientists already knew about from limit cycles, to the centre stage of the physics of fluid dynamics. Lorenz switched from mathematics to meteorology as a result of a wartime experience as a forecaster. In the early 1960s he built a 'toy weather machine' that could turn out daily weather records and had 'weather' of its own that could roughly imitate the real world in a passingly satisfactory way. It could show in early computer print-outs the pattern of the prevailing wind and the rotation of cyclones.[40]

One day in 1961, to save time, Lorenz typed in some duplicate numbers from a previous print-out in order to pick up on a programme that was half-way through its run. His new figures went only to three decimal places rather than the usual six. Lorenz was in for a big surprise. As the programme neared its new run he saw weather patterns diverging so rapidly from the previous run that 'he might as well have chosen two random weathers out of a hat'.[41] In a short time the tiny discrepancies had fed on themselves to become truly momentous. Lorenz's machine was repeating weather patterns over and over again, with very subtle variations that ultimately became very large ones.

Figure 1 on previous page shows Lorenz's 'butterfly wings' of two similar climate trajectories on the O-Z axis. One butterfly wing could represent a zonal regime, and the other a blocking high. In Figure 2 we see the weather states start to evolve. They are similar at the beginning (2a), and both are on the left side of the O-Z trajectory. There are now three possibilities of Lorenz's x, y and z: they all remain on the left wing (2a); both move over to the right wing (2b); just one moves over to the right wing (2c); In Figure 3a the weather states evolve similarly; in 3b they evolve differently.

Source: New Scientist

The 'Butterfly Effect'

Here was the true genesis of the Butterfly Effect — the celebrated concept of a butterfly flapping its wings in the jungles of Borneo causing a storm in New York. Errors and uncertainties would multiply helplessly like a cancerous growth, cascading upward through a chain of turbulent features, from sand storms and eddies up to continent-sized ocean gyres that only satellites can see.

At first Lorenz suspected computer problems, but soon discovered that the effect was related instead to the high sensitivity of his equations to the initial data. Lorenz knew only too well that fluid motions are governed by nonlinear equations that were highly sensitive to small changes. Weather predictions would be well nigh impossible — because far from being a mere mathematical novelty, chaos existed in the real atmosphere too. In other words it would be impossible to predict the precise state of even simple dynamic systems like the weather because if all of a weather event's subtle starting conditions are known it may be folly even to attempt five-to-ten-day forecasts.[42]

Lorenz had to make the system more sensitive to small changes which necessitated the building of a 'model climate' with just three variables — x, y and z. One butterfly wing could represent a zonal regime (fig. 1a), and the other, say, a blocking high. Now the two weather states start to evolve. They are similar at the beginning, both on the left side of the O-Z trajectory. There now exist three possibilities of Lorenz's x, y and z: 1) both trajectories remain on the left wing; 2) both move over to the right wing; or 3) just one moves over to the right wing.[43]

But in all three cases the two trajectories have diverged, implying quite different forecasts. On the other hand in the first two scenarios (2a and 2b) the two trajectories evolve along similar lines, from unsettled to more settled. This means then that only in the early stages can you make predictions about the weather. From the second scenario of 2b onwards the situation is getting a little out of control, as it is evolving quite differently. So we can see from fig. 3 that we can give quite a good forecast in figure 3a, but not with confidence in figure 3b because of the wide meanderings away from the initial state.

The Lorenz attractor implied there is no pre-ordained number of times a given trajectory must circle around one of the butterfly wings; it could be once or any number of times, depending on the nature of the trajectory. It is still not known whether the climate attractor will suffer from minor perturbations, or whether the entire shape and position of the attractor will be dramatically changed, leading to devastating weather states not experienced today. Unusual runs of unseasonal weather might simply be a manifestation of the small part of the vector field that for no apparent reason has lurched towards another part of the field associated with drastically different weather modes — instead of a hot summer we get a cool, wet one.

Simple convection models effectively demonstrate that the weather will follow broadly similar patterns which define the climate, but will never return to the same mode in the same way that climate can flip randomly between stable states in an unpredictable manner. The collapse of vast ice sheets in the Antarctic would be such an event.

On the other hand, fluid dynamics and solar radiation are the two features that give climate its order, if it is to have any order. This, in colloquial terms, means that London will *never* get deep snow in July, even if it gets cool rains in June. Even in the mildest English winter the temperature cannot reach 80F. Hence, chaos theory is circumscribed by the inherent order of large-scale physical systems obeying largely Newtonian laws. Again, greater accuracy can occur in forecasts for settled Mediterranean climes where atmospheric chaos is not so pronounced as it is, say, over moisture-laden Britain. In actual fact weather forecasts today are more accurate than they were some 15 years ago because the number of degrees of freedom away from the Lorenz 3-variables have been built in.

In any event there is the annual cycle of the seasons which gives order, and is surprisingly the one feature that climatologists are loath to look at. Professor Sir Brian Pippard, a physicist at Cambridge, says for this reason we should not be so harsh about chaos, since it will not prevent us from predicting future climatic change. Sir Brian says we can make confident predictions over long time-spans, and says the task is to discover the range of predictability in each chaotic system we are studying. He admits, though, that it is not easy.[45]

Nevertheless, many scientists believe chaos promises to bring discipline to the fuzzy boundary between order and disorder in the natural world. The Americans have even made an academic virtue out of chaos. There are now chaos journals and chaos conferences, even involving the CIA and the Dept. of Energy. There are students who make their first allegiance to chaos and their second to their specialisms, but cynics might argue that that is the natural order of things. At the University of Los Alamos there is now a Centre for Nonlinear Studies to oversee work on chaos and its related problems.[46]

FOOTNOTE REFERENCES:
Chapter Six: Can Climatologists Predict?

1. *Searching for Certainty*, John L. Casti, Abacus, 1993, p26.
2. New Scientist, book review, 25/5/91.
3. New Scientist, Paul Simons, 7/11/92.
4. New Scientist 13/4/94.
5. New Scientist, Ian Strangeways, 23/11/91.
6. Ibid.
7. Ibid.
8. Ibid.
9. Ibid.
10. *Our Drowning World*, Prism Press, 1989, p28.
11. Ibid, p26.
12. New Scientist, op cit.
13. Scientific American, August 1990.
14. New Scientist, op cit, 23/11/91.
15. Discover, (U.S.) October 1992.
16. Times Higher Education Supplement, 21/2/92.
17. Geographical Magazine, July 1989.
18. Searching for Certainty, op cit, p94.
19. Ibid, p96.
20. Ibid, p114.
21. The Times, 22/2/90.
22. Sunday Times, 4/11/90.
23. The Times, 9/3/94.
24. Ibid.
25. New Scientist, 25/5/91, Mike Hulme, book review.
26. *Living in the Greenhouse*, Michael Allaby, Thorsons, 1989, p121.
27. New Scientist, Paul Simons, 7/11/92.
28. New Scientist, 23/11/91.
29. Discover (US), November 1993.
30. Scientific American, August 1990.
31. Ibid.
32. Ibid.
33. New Scientist, 23/11/91.

34. Nature, vol 345, 14/6/90.
35. Scientic American, July 1990.
36. New Scientist, Tim Palmer, 11/11/89.
37. Sunday Times, op cit, 4/11/90.
38. Ibid.
39. Searching for Certainty, ibid, p56.
40. The Times, 4/8/90.
41. The Times, op cit, 29/10/90.
42. New Scientist, op cit, 11/11/89.
43. Ibid.
44. The Times, letters, 2/5/91.
45. The Times, op cit, 4/8/90.

Chapter Seven:

THE COSMIC CONNECTION

Postwar environmentalism is highly anomalous. It decrees that Man is above and more powerful than Nature, and unless some of his more profligate and destructive activities are severely curtailed he will sow the seeds of destruction in his own corner of the universe.

It is worth stressing how recent and unusual is this anthropogenic perspective, so contrary is it to hundreds of years of a kind of deism that has genuflected to the fearsome characteristics of the sun. For the myths and legends of ancient civilization acknowledged the vulnerability of the Earth and its people to cosmic and solar phenomena. Many pre-literate communities have, for millennia, worshipped the sun as a god, realising how dependent they were upon it, and some writers suggest that sun-worship is at the root of most religious festivals.

In some early catastrophe theories a wildly careering sun was blamed for causing countless Earthly disasters. In one story by Plato we read how Phaethon, whose father was the sun, lost control of his father's chariot to bring about Earthly doom. Egyptian priests told Herodotus that 11,000 years before the axis of the Earth became displaced, 'the sun had removed from his proper course four times and had risen where he now setteth and set where he now riseth.'

Nowadays the concern about the significance of *sunspots* might well imply that the solar deism of the ancients has merely donned a more reasoned, scientific, persona. A growing number of scientists, and many amateurs, think they have detected virtually indisputable correlations between sunspots and climate.

Robert Jastrow, an astrophysicist at Dartmouth College, New Hampshire, said solar activity had been 'low' in the 13th, 15th, 17th and 18th centuries and that if correlations with the past persist the 21st century will be a cool one. Jastrow, like others, believes that the present century has shown no clear trend, but its checkered climatic career nevertheless has something to do with sunspots.

Sunspot theorists say the famous half-a-degree rise in the temperature this century may have something to do with a fluctuating sun. It probably also accounts for the cooling from between 1940 and 1970 which science writer Nigel Calder reminds us makes no sense at all in regard to the greenhouse warming argument.[1] Michael Schlesinger of the University of Illinois at Urbana-Champaign also says that if part of the warming this century has been due to changes in solar activity, then 'the additional greenhouse effect may be weaker than was previously thought'.[2]

Correlations, whether causative or not, abound. Joe King, a physicist and former employee at the Radio & Space Research Station in Slough, and a long-standing advocate of the cosmic connection, says he has found correlations between cricket scores and sunspots, even publishing a scientific paper about it in *Nature* in 1973.[3] In 1982 Karin Labitzke of the Free University in Berlin received widespread publicity in the scientific media when he said there seemed to be a link between sunspots and stratospheric events at the equator and at the polar regions, and that he was able to make accurate weather predictions as a result.

The drought in the Sahel, which ended in catastrophic floods in 1988, was also said to have occurred when the sunspot cycle (of which more later) was at its maxima. Scientists at America's NOAA once suggested that droughts in the American midwest seem to occur roughly a year or two after a double sunspot cycle.[4]

In 1990 it was reported that Britain's Met Office had formulated a new long-range monthly weather forecast, to complement its Weatherplan for industrial customers, by relating idiosyncratic solar activity to weather changes.[5] These forecasts were produced on a personal computer by Piers Corbyn, a former astrophysicist at London's South Bank Polytechnic. Corbyn has had some forecasting success, actually winning thousands of pounds from bookmakers. The Met Office said Corbyn's forecasts were spot-on for seven out of ten months during 1989/90.[6] Corbyn successfully predicted the unusual weather events that took place in Britain in 1990. His results, he said, were better than chance. He believes that the solar connection can be proved by matching up unusual solar activity with past weather patterns, using weather and astronomical records dating back to the 17th century.

Traditionally, it was thought that an absence of spots on the sun coincided with colder weather. It was well known that during the cold period in the northern hemisphere between 1640 and 1710 the number of sunspots was very low: it became known as the 'Maunder Minimum' after the scientist who first perceived the correlation between sunspots and 'mini ice ages'. From 1710 onwards the sun was supposed to have 'switched (back) on'.[7] A similar period of low solar activity and low temperatures, called the Sporer Minimum, took place between 1400 and 1510.[8]

Eigil Friis-Christensen and Knud Lassen of the Danish Meteorological Institute show how well the correlation worked from 1750 onwards, closely following an 11-year sunspot cycle. They said that when the cycle is at its longest the decline in Earth temperatures are so good a fit, the chance of them being a fluke is one in 20.[9] They used figures on sunspots supplied by the Zurich Observatory together with statistics on global warming published by the UN Committee on Climate (the IPCC). Their graph shows that after a dip in about 1895 both sunspot activity and global temperatures rose steadily until about 1940, falling again until 1970, and rising thereafter.

Northern Ireland is an interesting corner of the world where untidy heaps of manuscript records and other more modern hi-tech data produce intriguing sunspot/climate conclusions. At Armagh Observatory the records of air temperature go back to the end of the 18th century, and the sunspot data go back before the Industrial Revolution. Together they virtually account for all of the climate's vagaries, as well as the short-lived effects of volcanoes.[10] These findings were confirmed at the European and National Astronomy meeting in Edinburgh in April 1994. One speaker said he found a close correlation between temperatures measured at Armagh since 1795 and the length of the sunspot cycle.[11]

There is much tangible evidence of historic sunspots. The level of cosmic rays from the sun can be recorded in tree rings, and degrees of magnetic activity are also revealed in radioactive carbon-14. Trace elements are as important as periodicity in proving the sunspot connection. In the early '80s Edward Zeller of Kansas University traced nitrogen compound samples of polar ice. He reasoned that it could only have been solar or cosmic events that could have dissociated nitrogen particles from the atmosphere. Nitrogen-dating techniques can now match up ice core readings with historical weather recordings.[12]

Rocks in southern Australia dating back hundreds of millions of years also contain strata showing 11 and 22 year rhythms. Researchers at Nasa's Ames Research Centre in California report a varve periodicity of 10.8 years.[13] Evidence from ice cores drilled in Greenland also suggest variations in northern temperatures over the past thousand years have occurred in cycles of about 80 years.[14]

Why Sunspots Occur

The physics that explain the existence of sunspots apply, of course, to all stars. Like the others, the sun is a huge sphere of incandescent gas, and generates heat and light in its core by nuclear reaction in which atoms of the lightest element, hydrogen, combine to form atoms of the next lightest, helium. Though the core of the sun is fantastically dense and hot, some millions of degrees Centigrade, the sunlight that we see and feel emanates from the surrounding atomic structures, and is released and re-absorbed millions of times. Indeed it takes so long for the light to escape from the sun that today's sunlight was first generated in the year 8,000 BC. The shining surface of the sun, the photosphere, has a temperature of a mere 8,000C, say some scientists. Yet its corona, its outer atmosphere, according to researchers at the European Space Agency, is heated to over 1m degrees Centigrade.[15]

All of the sun's energy received at the Earth's surface is a variety of electromagnetism (EM), so one form of EM on a different part of the spectrum can increase relative to other parts. This is an important point, since many of the mysteries concerning the impact of the sun on our climate relate to the part of the sun we are talking about. Even today there is still a certain vagueness about this, giving much ammunition to the Greenhouse theorists.

Clearly the nuclear furnace at the core of the sun, being to some extent 'organic', does not function with a machine-like or mathematical regularity. But it can abruptly yield more radiative energy from time to time. Some scientists believe this happens because the sun fluctuates, like a thermostat, reacting to conditions in the surrounding galaxy. Ronald Gilliland, of NCAR, says the sun appears to breathe in and out every 76 years, as confirmed by historical records of its diameter.[16]

All stars seem to be shot through with huge loops of magnetic energy generated deep within their cores. At the surface the sun seems to hum and vibrate, enabling astronomers to peer into its depths, rather as seismic waves hint at what is going on below the mantle of the Earth. Every now and then the loops snap to create discharges of hot gas and particles. They smash into the polar regions of the Earth to create auroral displays and disrupt our own magnetic field. The sun, in effect, shoots tiny bits of itself into space, namely hydrogen nuclei (protons) and negatively charged electrons. But they can be flung out in three different ways. First there are the flares which often combine with much brighter faculae (torches). Sometimes 10 flares a day can be observed reaching markedly higher temperatures than that at the surface — up to above 20,000F. Then there is the solar 'wind', only discovered in the 1940s, which is a continuous outpouring of high-energy particles from the corona at temperatures, as we have seen, of more than a million degrees.

But it is the curious dark blotches on the sun — the sunspots — that stimulate the most interest. They have been suspected for a long time of having some connection with a star's variable output. The Mount Wilson Observatory showed in 1994 that other sun-sized stars also have sunspots that disperse energy unevenly and reveal cyclical activity.[17] This confirmed what Sallie Baliunas of the Harvard-Smithsonian Centre for Astrophysics and Robert Jastrow of Dartmouth College proved in 1990, after spectroscopically analyzing stars going back to the mid 1960s.[18]

The Birmingham Solar Seismology Network have found that the sun's magnetic vibrations weaken when the sun is most active. Magnetic pulses associated with sunspots seem to impede the flow of the nuclear chain reactions that gradually make their way to the surface, so that when sunspots do break out the surface gets hotter.[19]

The darkness of the sunspot, however, is illusory, simply appearing dark against the bright glowing surface of the sun's photosphere. In fact the variation in brightness is insignificant: it varies by about a tenth of 1 per cent during an 11-year cycle, according to Jastrow. But it could change upwards now and then. The cycle itself is due to the rotating nature of the sun's field.

The size of the spots on our sun seem to change frequently. Although many are only a few hundred miles across, some can range up to 10,000 miles in diameter. There is an unmistakable connection between sunspots and flares, and the two phenomena often appear together. They form in groups, often dominated by a prominent pair which seem to grow cell-like until they absorb the smaller spots. Some spots can last months, and others only a few hours. But the cycle is by no means smooth and clear-cut. Peaks in intensity vary from cycle to cycle, and the peaks themselves seem to have been more intense over the past 60 years.[20] During the past 100 years the actual cycle has been reduced from 11.7 years to 9.7 years, but the cycle can run anywhere from nine to 13 years.

The number of sunspots alleged to have been observed in historical epochs has sometimes been disputed. For example the 19th century Swiss astronomer Rudolf Wolf collected thousands of records of past solar and sunspot events, now known as the Wolf sunspot number. But a group of American and French astronomers used records from the Royal Greenwich Observatory and Greek and other records dating back to the early 1600s, amounting to well over a third of a million observations, and found fewer sunspots than Wolf did.[21]

Generally speaking, there is a fairly clear-cut common-sense relationship between brightness and solar activity. The satellite Solar Maximum Mission (SMM) proved this in 1988.[22] Between late 1980 (just after the peak of the last solar cycle) and mid 1985 solar activity declined,[23] and coincided with a steady decrease in the sun's luminoscity. According to the Rutherford-Appleton Laboratory at Harwell, Oxfordshire, 1990 was a maximum year for sunspot activity,[24] and this too coincided with a warm summer in Britain.

Let me repeat: most of the turbulence on the sun arises from its magnetic field. Using data dating back to 1981 from a high altitude observatory in the former Soviet Union, it was found that the sun during its most active phase, is accompanied by strong bursts of radio noise.[25] The sun emits radiation on a variety of spectra, and light and heat are just one. But it also gives off X-rays, UV light, speeding protons, gamma rays and radio waves. 'There's a lot of energy being dumped into the Earth's atmosphere by the sun, through things like flares', pointed out Bill Stuart, head of the Geomagnetism Research Unit at the British Geological Survey.[26] Stuart believes that solar storms may thus affect the weather through the butterfly effect.

Fast-moving charged particles heading towards Earth do seem alarming by themselves. Entire doomsday scenarios, including the death of the dinosaurs, have been constructed around such potentially terrifying events. In recent times sudden solar discharges hurled towards the planet are sometimes picked up as electric storms that can cause electricity blackouts.[27]

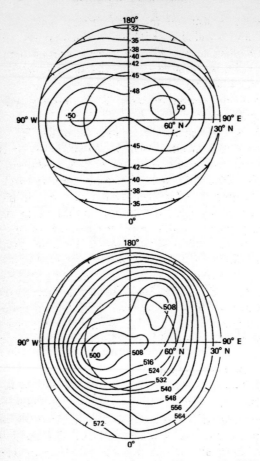

The top illustration shows Earth's magnetic field for the northern hemisphere for 1965. The atmospheric pressure system for the same period shows remarkable similarities.

Source: John Gribbin, Forecasts, Famines & Freezes, Wildwood House, 1976, after J.W. King, Nature, October 1973

The search for an explanation of magnetic (rather than electric) storms, most frequent during periods of the greatest solar activity, has intensified since a severe one in 1989 deprived six million people of electricity in Quebec and disabled a nuclear power station.[28] In that year the British Geological Survey put out a warning of severe magnetic storms for June 18 because of the high number of flares seen on the sun's surface, which may actually have been the so-called coronal holes which appear only as dark patches in X-ray images of the sun. In any event nothing untoward happened on Earth. In fact we should be thankful for the aesthetic pleasure some of the flares give us. The aurora borealis at the north pole are a manifestation of such, as the protons interact with Earth's magnetic field.

The sun is still capable of yielding surprises. The spacecraft Ulysses passed underneath the south pole of the sun in September 1994 after a four-year trip which took it on a 'slingshot' journey around Jupiter. Robert Forsythe, a member of the research team that launched the probe in October 1990 under the auspices of the European Space Agency, said he was surprised to discover an unchanging magnetic reading from the 'bottom' of the sun. Maps of the star's surface, derived from observations on Earth, show it is a magnetic dipole, and computer models suggest that the magnetic field should be stronger over the sun's poles.[29] Theories evolved to explain the sun's surface field will therefore need revising, he said. He added that this would also raise questions about the true extent of our knowledge of the solar wind, while the new science of magneto-meteorology is still in its infancy.

The Magnetic Explanation

Goesta Wollin, of the Lamont-Doherty Observatory, once suggested that the weather/solar connection is the linkage between the magnetic fields of both the sun and the Earth.[30] This is because the atmosphere also has a magnetic field, like the Earth itself. Magnetism, after all, is one version of electromagnetism, the invisible force field that permeates the universe. In addition the sun's field must be affected by its own orbital spin. There is some evidence that the spin at its equator sped up some five per cent during the Little Ice Age, all the while carrying the sunspots with it.

The atmosphere, like all gases, must consist of billions of sub-atomic particles. Over 100 years ago, one Balfour Stewart suggested that not only can a current flow through the gases of the atmosphere, but with its hurricanes and storms it can also generate kinetic energy. Of course it is not easy to take direct measurements of electric energy in the atmosphere, but they can be worked out from levels of conductivity and electric field data by the use of Ohm's Law. The knock-on effect largely depends on the current state of geomagnetic play — whether the Earth's field is increasing or decreasing, and in what way, and in which hemispheres. Unfortunately the field tends to vary from one part of the world to another, and from one epoch to another.

Sunspots do seem to coincide with periods of global warming wherever Earth's field has weakened, or is not particularly pronounced, or has drifted somehow out of alignment. Here we have introduced the subject of *periodicity* into meteorological factors, and as mentioned in Chapter Five, it is scientifically unlikely that the atmosphere has its own built-in 11-year fluctuations. One would rather accept the astronomical explanation. Perhaps two natural fluctuations are interacting with each other, thus giving statistical improbabilities slightly better odds. The danger seems to be when the field, as it has done from time to time throughout history, occasionally 'fails', to allow lethal protons to enter the atmosphere and to re-arrange the particles therein.

In short, changes in atmospheric chemistry could in turn affect Earth's radiative balance, although unlike the CO^2 radiation theories they would act through electrical rather than radiative principles. To understand how this might work let us start with the highest point in the upper atmosphere — the ozone layer, some ten miles up. Remember that the atomic nuclei the sun ejects violently into space are a form of cosmic ray — the solar wind. Cosmic rays themselves could be funnelled round the poles where the magnetic dipoles operate. This in turn could affect the chemistry of the atmosphere, especially the reactive chemicals therein such as 'free radicals'. It was once suggested that occasionally the sun passed through an interstellar cloud of hydrogen and helium particles. On reaching a certain density the hydrogen cloud could reach Earth's environment and react with the hydroxyl 'radicals' (HO), to turn them to water vapour. Climatologist Walter Orr Roberts believes solar ray bursts could cause low pressure systems in the northern Pacific.[31]

Yet in penetrating the atmosphere to reach the ground, cosmic rays collide with molecules in the air layers to produce electron 'showers'. These secondary electrons in turn often ionize molecules to form negative ions. A typical multiplication process takes place, rather like the way a few chlorine atoms can destroy tens of thousands of ozone molecules. One cosmic particle can create as many as one billion ion pairs. Some scientists point to the connection between phosphorus-32 and beryllium-7, two important elements that move down from the stratosphere to the troposphere within 48 hours of a solar flare occurring.

The ionosphere can actually be dragged via fierce winds across Earth's magnetic field to produce a primitive voltage. This was scientifically confirmed by Earle Williams of the MIT in 1992.[32] The atmosphere behaves like a spherical capacitor, with the solid Earth below acting as a lightning conductor. Colin Price of Columbia University suggests that between the two is a 'leaky dielectric'. Price has confirmed the ionosphere's potential, and with the aid of satellite data has worked out that a 1C rise in average surface temperature increases the electric potential by 10 kilovolts.[33] When the sun flares up more than usual there is a knock-on effect in the atmosphere similar to ionisation, but instead of atoms being torn apart the larger molecules of gas are split into atoms. This tends to reduce ozone levels because the pulverized oxygen atoms are prevented from assuming their threesome state, and become instead ordinary molecules of oxygen. Computer models show that ozone does decrease slightly over the 11-year cycle, and the US-German Sapex satellite recently launched suggested that electrons from the sun are to blame for ozone-thinning.[34] Some scientists believe the 1930s were warm for this reason.

I dismissed, in an earlier chapter, the argument that depleted ozone levels have much to do with climatic change. But it is worth stressing here that some scientists suggest that a diminished ozone layer can interfere with the delicate oxygen/nitrogen balance in the atmosphere by creating nitric oxide (NO), or other oxides of nitrogen (collectively known as NOx). Bear in mind that only about one hundredth of the sun's rays are in the UV band, which is a higher frequency than other spectra. But this percentage greatly increases with observed *changes* in solar output[35] because the extra UV comes from the higher temperature region near the sun's surface. Scientists from Imperial College, London, have suggested that any extra UV would have a disproportionate effect on the stratosphere.[36]

Furthermore, the water vapour in the stratosphere and troposphere below the upper ozone layer may well play a decisive role, especially in conjunction with the magnetic field. The field could drive the electric part of the weather machine.

Maybe the link is to do with the world's electrostatically conductive oceans. Or perhaps the evaporation of water from the surface negatively charges the dry earth, and positively charges water vapour. In fact Volta, after whom the volt was named, in 1800 advanced an electrification theory based on vapourization of water to explain thunder. Indeed thunderstorms must obviously have something to do with charging up the atmosphere. They undoubtedly have the ability to dissipate atmospheric heat; people notice how much cooler it has become after a storm has 'cleared the air'. Other electro-chemical features may be involved with negative ions perhaps being captured by the land surface through some electrostasis phenomenon.

It is as well to recall that the planet has two hemispheric weather systems circulating in the north and south in a kind of dumb-bell configuration. Ridges of high pressure often accompany any slight slackening of magnetic intensity in the Earth's northern force field. Storms and blizzards in America might be due to the sudden changes in the strength of the solar magnetic field, and successful weather predictions have been based on this hypothesis.

Perhaps the electrically charged particles from sunspots affect the westerlies and the jet streams, suggested Joe King, whom we met earlier.[37] The winds blow above the equator and reverse their direction every 12 to 15 months.[38] The westerlies, in addition, seem to coincide with periods of high solar activity — corresponding with the 11-year sunspot cycle. This link was first highlighted by meteorologists Harry van Loon and Karin Labitzke, of the Free University in Berlin.

So perhaps the scientists are telling us that a longer cosmic cycle could produce shorter terrestrial ones. El Nino is cyclical, but, as we saw in Chapter five, no-one could put a reason to it. As early as 1971 scientists from NCAR discovered short-term oscillations in winds in the central Pacific region, and this was later confirmed by geostationary weather satellites. Hence a tentative link was made — tropospheric winds creating unusual cloud patterns seemed to occur over the Indian Ocean every 40 or 50 days,[39] and seemed to produce more blocking events in mid-latitudes.

The Cosmic Dissenters
The greenhouse-supporters deny the solar thesis. The chief difficulty recognised by sunspot enthusiasts is that periods of both warming and cooling have at times coincided with flares and sunspots. Critics at the CRU, such as Philip Jones, say the sunspot-climate link is never satisfactorily explained,[40] and that satellite

measurements prove little. The more mini-links that are postulated, via cosmic rays, ionisation and stratospheric winds, for example, the more likely the overall explanation for the links will elude us.

The successes of sunspot/climate predictors are seen by their opponents as transient. For example, in 1988 there was a high expectation that as the winds turned westerly as solar activity increased, the US winter of 1988/89 would be severe. However, the prediction did not come to fruition.[41]

Anthony Barnston and Robert Livezey from the Climatic Analysis Centre of the US National Weather Service, using the sunspot basis, also attempted to predict a severe US winter in 1988/89. But they admitted their failure might have been due to the fact that they didn't take El Nino into account. But as we have seen, some scientists believe El Nino is also due to cosmic factors. The opposite situation to El Nino is its gender opposite — La Nina — which is also said to be a function of unusual solar factors.[42] La Nina was thought to have cancelled out otherwise severe weather that El Nino alone could have produced.

And whereas critics of the greenhouse theories say that the alleged warming has been minuscule, the greenhouse-mongers themselves turn the argument on its head. Tom Wigley and Sarah Raper of the CRU say that solar changes since 1874 would only account for an unobservable fraction of a change in temperature, something like 0.05C. Nevertheless, the greenhouse advocates, like others at the CRU, by arguing that the world has warmed by ten times as much since, leave themselves open to the criticism that as an empirical fact this has not yet been proved. Yet they have a valid point when they say that the energy variations in the sun would be too small to explain long-term climatic changes, although one suspects that they are referring merely to infra-red energy. One scientist from the Hadley Centre was quoted as saying that the 11-year cycle was measured by satellites to change the 'energy input' at the top of the atmosphere by a mere 0.2 watts per square meter.[43]

There is also the problem of daily time-lags with changes in solar radiation being absorbed via the 'weather machine' in the troposphere. Another drawback with the linkage is the lack of forecasting certitude that one could get with more conventional meteorological tools. Even so, Nigel Calder suggests the IPCC has relegated solar inputs to the bottom of the agenda because of inadequate spacecraft readings of the sun's output. He says the greenhouse bandwaggon has too much political momentum to be stopped now.[44]

Calder may not be quite right about this. The warming scenario was a 1980s message, and times have moved on since then. Many climatologists of all persuasions are saying openly that we need to focus more on the behaviour of the sun in all future discussions about the climate. David Rind of the Goddard Space Institute admits that our 'ignorance is almost total' in regard to the way the sun affects Earth's climate. To try to remedy this the Upper Atmosphere Research satellite was launched in September 1991, and is to be backed up in 1998 by the Earth Observing System satellite which will detect how the 'solar constant' changes over time.[45]

The European Space Agency will, in 1995, start monitoring solar seismology in detail with its Solar and Heliospheric Observatory, SOHO, and Cluster space missions.[46] Britain has contributed some £70 million to this venture. The SOHO will lurk between the sun and Earth. On board, a British-built instrument will try to find out why the sun's outer atmosphere is heated to over one million degrees. The solar wind and its link-up with Earth's magnetic field will be measured by Cluster, a joint ESA/Nasa project, using four spacecraft to catch the shifting nature of charged electrons.

Many are convinced that this latter project will finally prove that this curious electromagnetic linkage will explain virtually all of Earth's strange weather anomalies. And, like our early ancestors, we may become solar deists.

FOOTNOTE REFERENCES:
Chapter Seven: The Cosmic Connection:
1. The Guardian, 21/4/94.
2. Nature, vol 360, p330, December 1992.
3. Sunday Times, 13/5/90.
4. *Earth's Changing Climate*, Antony Milne, Prism Press, 1989, p57.
5. New Scientist, 23/6/90.
6. The Times, 9/8/90.
7. New Scientist, 29/10/94.
8. Astronomy, August 1988.
9. Daily Telegraph 24/11/91.
10. Guardian, op cit, 21/4/94.
11. The Times, 7/4/94.

12. *Our Drowning World*, Antony Milne, Prism Press, 1989, p99.
13. Earth's Changing climate, op cit, p56.
14. Science, vol 254, p698, 1991.
15. Daily Telegraph, 13/4/92.
16. *Our Drowning World*, op cit, p100.
17. Guardian, op cit, 21/4/94.
18. Nature, vol 348, p420, 1990.
19. Guardian, op cit, 21/4/94.
20. Geophysical Research Letters, vol 17, p2169, 1991.
21. Geophysical Research Letters, vol 21, p2067, 1994.
22. Nature, 28/4/88.
23. Astronomy, August 1988.
24. Sunday Times, op cit, 13/5/90.
25. New Scientist, 17/9/94.
26. Daily Telegraph, op cit, 31/12/90.
27. The Times, 3/10/94.
28. Sunday Times, 2/10/94.
29. The Times, op cit, 3/10/94.
30. Earth's Changing Climate, op cit, p58.
31. Ibid, p55.
32. Science, 13/6/92.
33. Geophysical Research Letters, vol 20, p1363, 1992.
34. Guardian, op cit, 21/4/94.
35. New Scientist, 5/1/91.
36. Nature, vol 370, September 1994,, p544.
37. The Times, op cit, 8/11/90.
38. New Scientist, op cit, 5/1/91.
39. Ibid.
40. Daily Telegraph, 24/11/91.
41. The Times, op cit, 8/11/90).
42. New Scientist, op cit, 5/1/91.
43. Daily Telegraph, 24/11/91).
44. Guardian, op cit, 21/4/94.
45. Daily Telegraph, 10/2/92.
46. Daily Telegraph, op cit, 13/4/92.

EPILOGUE

Has climate become a political issue?

The theme of this book has been the complexity of the subject of climate analysis. There is still much about it we do not know. It is the product of many complicated interractions between the oceans, the atmosphere and the uneven topography of the hemispheric land masses, all driven by energy from the sun, which itself may go through more fluctuative periods than hitherto thought.

The warming debate was renewed in early 1995 after the widespread floods that affected parts of California, and which had brought widespread devastation to Europe (said by the French to be the worst floods of the century). The climatologist from the National Weather Bureau in Maryland, America, put the cause down to a 'funnel' of atmospheric turbulence that involved Arctic winds and El Nino type events similar to that described in Chapter Five, but which also had knock-on effects in Europe. A British weather forecaster, however, said that the floods were simply due to Atlantic depressions that one would expect in winter, but with the added complication that areas of high pressure usually found in Europe at the time, and which would normally squeeze the depressions further northwards away from the Continent, simply weren't there. When asked, inevitably, by the newsreader, whether the floods were caused by the greenhouse effect, the Met man said that one would need more than 100 years of similar weather before we could even come to a tentative conclusion on the matter.

The rather chastening knowledge of recent years about chaos theory emphasizes how difficult it is to make predictions about the climate, even if we possessed complete knowledge of it. For what chaos theory proves is that the reactive features of weather and climate have formidable powers of self-correction (leading to a sometimes alarming over-correction). The GRIP and GISP ice cores, showing how rapidly, in geological terms, the climate has reversed itself, is further proof of the difficulty in, perhaps one might say the fatuity of, making climatic predictions.

Richard Alley, a geo-scientist at Pennsylvania State University, reminded us as recently as March 1995 that until the end of the last Ice Age the Greenland ice sheet showed that climate often changed rapidly from one stable state to another, with the temperature varying by as much as 7C. Alley warns against assuming that rapid 'flips' are a thing of the past. The cores show that just 8200 years ago there was a sudden cooling of about 4C, which lasted about 200 years.[1]

This knowledge risks turning climatology into a kind of metaphysics. We can see the climatic crossroads just ahead, says Michael Allaby, in his *Living in the Greenhouse*, but we are unable to make out the lettering on the signposts.[2] A more telling point is that it is Man himself who is approaching the signpost, and as he advances ever closer he begins to make occultic gesticulations and genuflections. Then the signpost suddenly springs into focus: it 'collapses into reality' as the quantum physicist might put it. It tells him that not only is the Earth heading for a warming but, since he is taking an inordinate interest in it, he might well be the cause of it.

This train of thought highlights a new and rather worrying development. The pursuit of climatological knowledge is a scientific endeavour, and it should be free from outside pressures as are other strictly academic subjects. Unfortunately as Britain's Meteorological Office is now an environmental agency it is also subject to the influences of pressure groups that obliges its scientists to pay attention to other non-academic perspectives.

However, the problem with environmental subjects is that they are collectively known, in some more cynical quarters, as *environmentalism*, and environmentalism shows all the signs of becoming a new kind of pastoral religion,[3] with all the familiar catechisms associated with religions: we are damned for our excesses and denounced for our arrogance towards nature. The Gaia Peace Atlas, named, if you recall, after the Greek goddess of the Earth, published by Pan Books in Britain, had a forward by a former UN Secretary-General, who spoke in eschatological terms: 'We have perhaps 15 years left ... 5,000 sunrises and glimmering dusks ...'[4]

Fifteen years is a very short time span. But what was surprising was the fact that the central dogma of climatology — that oceanic inertia virtually precludes any such swift turn-around in climate — was on the verge of being abandoned. It was, for instance, echoed by leading climatologists like Stephen Schneider, of the National Centre for Atmospheric Research in America, who himself wrote in a climate book in the 1970s: 'A cooling trend has set in — perhaps one akin to the little ice age".[5] Yet it soon looked like Schneider and others were actually *campaigning* for a warming.

But this might, on reflection, not be such a bad thing for the science itself. We must remember that scientists are no different from other professional groups. If you are working in climatology a good example of stimulating interest and getting backing for it is to overstate the cause, but with the best of motives. After all the study of climate is an important and worthwhile goal — as important, if not more so, than the study of high-energy physics. It is a big science, and needs big computers and satellites, and plenty of money. In Britain in 1993 more than £200 million was spent on research into the subject.[6] Indeed, Roy Spencer, a Nasa Space Centre scientist in Huntsville, Alabama, takes it for granted that the need for funds influences scientific conclusions, saying that science 'benefits from some scary scenarios'.[7] Hence a new phrase, similar to that of the over-zealous and recently privatized dentist, seemed to creep into the climatologist's vocabulary: *the precautionary principle*.

The problem is that the 'what if' questions about the future are all open-ended. Their hypothetical nature begs that something be done now, just in case. The argument against waiting for more scientific proof of the greenhouse effect was that, as John Gribbin once put it, 'when the rains start to fail in North America, and there is no spare grain to rush to famine regions' it will be too late.[8]

Unfortunately an infinite number of 'what if' questions can be asked. What if the flow of American rivers, already fully utilized, were to go down by 20 per cent? What would be the consequences for US agriculture if the rainbelt went further north? What would be the consequences for energy conservation, for national parks and the natural habitat? What will happen to global food supplies, and will the recurring disputatious water shortage in the Middle East be exacerbated? Indeed disturbing, yet hypothetical, musings turned to mild panic when one looked out of the window. In America, said Robert M. White, president of the US National Academy of

Engineering and formerly chief of the US Weather Bureau, mounting official concern was catalyzed, he believes, by the disastrous drought in the summer of 1988, when the Mississippi fell so low that navigation over long stretches were impossible, and crops throughout the grain belt were devastated.[9]

From then onwards the politicians and the journalists, who reported a seemingly increasing number of weather disasters, (on both sides of the Atlantic, which seemed to be significant) saw eye to eye. But what is curious to note is that the climatologists also saw fit to fall into line. True, not all of them did, but enough did so to give enormous validity to the issue. And by then, of course, they had lost control of it. A team of 170 scientists who had been closely involved in climate studies for years, and who had been individually selected by the World Met Office, produced a definitive study pointing to a minimum rise of 2C for every doubling of CO_2.

As many as 70 of the world's leading meteorologists also assembled in Egham in May 1990 to agree the final report of their 18 month investigation into the greenhouse effect. The group was chaired by Sir John Houghton, formerly chief executive of Britain's Met Office. The report said, in part, that by the year 2020 'global mean temperatures will have risen 1.8C above the pre-industrial level . . .'.[10] In the same year a statement by the Union of Concerned Scientists urged action in the US to curb warming gases, and was signed by 52 Nobel laureates and more than 700 members of the National Academy of Sciences.[11]

But it was the United Nations that pushed the subject centre stage. The organization had been influenced by the heavyweight study that was prepared in 1979 at the request of Frank Press, now president of the NAS, who was President Carter's science adviser. It was also in 1979 that the World Meteorological Organization convened in Geneva the first World Climate Conference. The UN then created the IPCC — itself a product of green activist campaigners — and set in train the task of reporting on all the ramifications of the alleged greenhouse effect. Its final report, co-authored by or endorsed by many leaders in the fields of atmospheric chemistry, meteorology and climatology, was quite doom-laden, as we saw in Chapter One. It detailed 'unprecedently rapid' changes in global weather patterns during the 1980s, when glaciers apparently melted at a great rate, and suggested that already 80 per cent of the world's coastline is beginning to erode. The report had an immediate impact on the international media and on national politicians.[12]

In Britain the entire liberal establishment became involved, and these days that includes virtually everybody who has cultural power — the Church of England, governmental agencies, the media, pressure groups and health campaigners, a growing number of doctors and scientists, and politicians of every party. Indeed politicians jumped on the bandwagon because it is an issue which seems, on the face of it, to be non-partisan, and a consensus seemed to be emerging. Implementing anti-greenhouse policies, or at least hinting that one would do so at international gatherings, had tremendous low-cost advantages. In America so great was the consensual pressure that a plan by the White House to play down the threat of global warming in 1990 had to be scuppered after European ministers refused to play along.

This is not to say that critics of the excessive emotionalism about climate issues had disappeared into the woodwork. Much of the caution and scepticism arose from the fact that there was no immediate concern for human health and no evident manifestation of climate change. The Bush administration in the late 80s had been at odds with the environmentalists, and at an ecology conference convened by the Dutch in November 1989 they and the Japanese refused to sign a draft resolution to 'stabilise' pollutants by the end of the 1990s. Congressional hearings followed hearings. Both the atmospheric scientists and the more general environmental activists began to choose sides in the argument, and debated whether immediate policy action was justified.[13] Officialy inspired 'working groups' planned for 'worst case scenarios'. One group, chaired by the Soviets, assessed the impacts of warming changes on agriculture, ecosytems and the economy. Another group, chaired by the US, was to devise policies. However, one critic pointed out that as these groups worked in tandem rather than in sequence, inconsistencies crept in with temperature and sea-level rises being highly varied.[14]

Soon the first parting of the ways amongst the international community occurred. Separate statements were drafted in Washington by the Europeans and Americans on what to do about the greenhouse gases. The European Commission has recommended energy taxes that would raise the price of electricity by up to 20 per cent.[15]

In the meantime some influential researchers, like oceanographer William Neirenberg of the Scripps Oceanic Institute, who had been working on climate and CO^2 emissions for more than 20 years, believed that policy actions were beginning to outrun the scientific evidence. Richard S. Lindzen of the MIT and Jerome Namias of Scripps, both distinguished long-range weather experts, wrote a letter to President Bush urging that no action be taken. Three members of the NAS, including Frederick Seitz, its former president, penned the Marshall Institute report, questioning the science underlining mooted policy actions. They said that recent climatic fluctuations could be explained by variations in solar radiation, and indeed, if that was the case the solar evidence pointed to a coming Little Ice Age rather than a warming in the future. They recommended more mathematical modelling.[16] In the US, said Lindzen, there has been a disturbing tendency towards what is known as 'result orientation': to encourage officials to reach certain conclusions.

Indeed any mooted consensus among climatologists seemed to vanish on closer examination. Although events portrayed might sometimes have a ring of plausibility about them, they should not be taken as real predictions, said some. Dr George Maul, a researcher at the National Oceanic and Atmospheric Administration in Miami, pleaded for environmentalists to preach of things 'that are really firmly established, not on the edge of scientific argument'.[17] Maul said the answers to two major questions about the greenhouse effect — how much warming will there be, and what damage it will do — are far from certain. Some studies show the warming could be very beneficial to food production, said Robert Pease, a climatologist at the University of California, Riverside.[18]

Wilfred Beckerman, an Oxford economist, was critical of the main pressure of the campaigners — to reduce carbon dioxide emissions, especially in the newly industrializing nations. He said that many people in Britain would in any event welcome the IPCC's prediction of a 3C warming by the end of the next century, and pointed to the popularity of sub-tropical Australia as a favoured destination of British emigrants.[19] He said a scenario assuming a continued rise in real incomes at an average level of about 1.5 per cent per annum over the next 100 years would mean the economic costs involved in a loss of agricultural output or implementing costly remedies should not be too onerous.

Jeremy Leggett, a geologist and environmental campaigner, said that out of 400 polled scientists, only 113 replied to a questionnaire, and of those only 15 were willing to confirm that they thought man-made global warming was likely.[20] Disagreements were particularly pronounced in America, where the reliability of scientific models was questioned. Farmers were also angered by the frequent bluntness of the warming advocates. One critic complained that it is not the job of scientists to sell anything, but simply to share the discoveries of the universe with the public.

However as time passed the green message began to slip down the political agenda. Indeed by 1995 *New Scientist* was reporting a 'greenhouse backlash', and campaigners from the energy industry were trying to cash in on the growing uncertainty about climate issues that the press were giving more attention to. A US weather forecasting company called Accu-Weather expressed reservations both about the facts of global warming and about its mooted anthropogenic characteristics. Just three years earlier, however, it appeared to support the IPCC's view that a warming was imminent. The company's new reports were commissioned by a lobbying organization called the Global Climate Coalition that aims to help powerful energy producers and consumers, i.e. those most likely to suffer from controls on carbon dioxide emissions, to contribute to the climate debate.[21] However some American gas companies, who emit less CO_2 than other fossil fuels, have joined an opposing pro-environment organization, which means that the fossil fuel lobby is now split.

In the meantime the industrialized nations are being urged to reduce CO_2 emissions by 20 per cent on 1990 levels. The Opec countries, not convinced of the warming argument, fear that agreements made at the Climate Conference in Berlin in 1995 will damage their oil markets.[22]

What is clear is that a more modest consensus seems to have emerged, and was articulated by the Met Office in late 1994 in a report. The rate of warming could be as slow as 0.15C per decade largely because of the extra ingredient of sulphate cooling, previously overlooked (see Ch 3). Britain, the report said, could take 70 years to heat up by even a single degree, and is more likely to become wetter than to experience droughts.[23] In the meantime greenhouse sceptics believe the estimates will continue to fall as the complexities of the climate are better understood. As Richard Lindzen said: 'We will look back on present (greenhouse) proclamations as hopelessly naive'.[24]

In view of the uncertainty still surrounding the issue we may, of course, not think that at all.

FOOTNOTE REFERENCES:
EPILOGUE

1. Living in the Greenhouse, Michael Allaby, Thorsons, 1990, p184.
2. Sunday Times, 10/6/90.
3. Sunday Times, Hilary Lawson, 12/8/90.
4. Sunday Times, op cit, 10/6/90.
5. Sunday Times, 11/9/94.
6. Sunday Times, op cit, 12/8/90.
7. New Scientist, John Gribbin, 28/7/90.
8. Scientific American, July 1990.
9. The Times, 23/5/90.
10. Scientific American, op cit, July 1990.
11. Observer, 25/2/90.
12. Scientific American, op cit, July 1990.
13. New Scientist, Deborah MacKenzie, 23/6/90.
14. New Scientist, op cit, 11.9.94.
15. Scientific American, op cit, July 1990.
16. Sunday Times, op cit, 10/6/90.
17. Ibid.
18. The Times, 24/10/90/
19. Sunday Times, op cit, 10/6/90.
20. New Scientist, op cit, 15/12/90.
21. New Scientist, 25/3/95.
22. The Guardian, 25/3/95.
23. Sunday Times, op cit, 11/9/94.
24. Nature, vol 345, 14/6/90.

GLOSSARY

A Acoustic Tomography   The study of the temperature of the stratified layers of the ocean with the aid of loud sound waves. Cooler water transmits sound faster than fractionally warmer water. One drawback is that below about 1,000 feet the increased pressure of the water can offset these sound-transmitting characteristics.

A Atmosphere   The air surrounding the planet. It has no precise upper limit, although it is generally thought to be about 200 kilometres from the surface. The density of the atmosphere declines rapidly with height, and about three-quarters of the mass is contained within the lowest most important layer, the troposphere, whose depth varies between 10km and 17km. The mesophere extends from the top of the stratosphere to over 80km above the surface. The thermosphere extends from the mesophere to the atmosphere's outer limits.

The various layers are distinguished by differences in the rate at which temperature changes with height, which either favours or discourages convection. There is much less vertical motion in the stratosphere than in the troposphere.

B Biological Pump: An organic recycling phenomenon, where carbon is taken from the upper ocean layers by plant-like micro-organisms (phytoplankton) which are in turn consumed, on a seasonal basis, by zooplankton, with the zooplankton also recycling nutrients that the phytoplankton need. This phenomenon has an important bearing on carbon dioxide levels.

B Biomass: The Earth's total stock of living green matter, including the woodlands, forests, jungles, cereals, grasses and other vegetation.

B 'Butterfly Effect':   A classic term in chaos theory, emphasing how the smallest of perturbations in one part of the atmosphere can have devastating consequences elsewhere. It was once thought this was purely a mathematical novelty generated by computers, but is now thought to apply to the real world too. The implications are that it is impossible to predict the precise outcome of weather and climate events from initial conditions.

C Circumpolar anticyclones: A prolonged high pressure system connected to the behaviour of the jet streams that circulate around the polar regions about eight miles up, and which affect other wind systems such as the Westerlies. They are responsible for prolonged episodes of cold, dry weather in winter, or for long 'heat waves'. They are sometimes known as 'blocking highs', since they can pincer-out other, more moderating, lower pressure systems.

C 'Climatic flip': The 'sudden' reversal of a continuous trend in the climate, hitherto lasting tens of thousands of years, within a geologically short space time of time, sometimes within tens of years.

C Coriolis Effect (or Force): Apparent force acting on moving objects that results from Earth's rotation. It causes oceanic and atmospheric currents, including the gyres and the Gulf Stream (see below) to be deflected to the right in the northern hemisphere and to the left in the southern hemisphere. The force is proportional to the speed and latitude of the moving currents, and is more pronounced at the poles.

C 'Curve-fitting': A mathematical device used in forecasting procedures. Data is 'plotted' on a graph, fed into a computer, when then it generates equations which are used for predictions. Curves can be made to differ mathematically and can yield a variety of extrapolations about the future.

E El Nino: A warm-water current which periodically reverses direction to flow south-west along the coast of Ecuador around Christmas time. It is also known as the El Nino-Southern Oscillation (ENSO), and has climatic effects throughout the Pacific region, and sometimes beyond. It has an unexplained periodicity of about seven years, which seems to be decreasing.

F Feedbacks: The way natural biological or geophysical events can countervail a trend (a 'negative feedback') or reinforce a trend, sometimes exponentially (a 'positive feedback'). In nature negative feedbacks are far more common, otherwise most natural and evolutionary processes on Earth would soon come to a halt.

F Fluid Dynamics: A study of the physical characteristics, properties and behaviour of fluids, including gases. Fluid dynamics is an important discipline in meteorology, because of the turbulent 'fluid' nature of weather processes.

G   Geostationary orbit: A satellite that remains virtually fixed in its orbit over one portion of the globe.

G   General Circulation Model (GCM):   A computerized 3-D model of the atmosphere as it is understood by climatologists, with the globe represented by tens of thousands of grid cubes, into which numerical data from weather stations and other more theoretical assumptions are fed. Climate projections are then made after further mathematical work is done.

G   Greenhouse Effect: A widely-used but somewhat flawed term, since it engenders a wrong image of how the climate works. Theoretically there is a possibility of heat retention in the lower atmosphere as a result of absorption and reradiation by clouds and gases, including water vapour, CO^2, methane and chlorofluorocarbons. The gases are transparent to incoming short-wave radiation, but partly opaque to infra-red radiation. However any warming brought about in this way is greatly countermanded by latent heat transfer, wind and weather systems, and additional cooling pollutants.

G   Gulf Stream: The most important ocean-current climate-shaping system in the northern hemisphere, stretching from Florida to north-western Europe. It incorporates several other currents, such as the North Atlantic Drift and the Florida current. See 'gyres' and SSTs below.

G   Gyres: Circular or spiral motion of water, usually applied by scientists to a semiclosed current system. A major gyre exists in each of the main ocean basins, centred at about 30 degrees from the equator and displaced towards the western sides of the ocean. Gyres are generated by surface winds and travel counter-clockwise in the northern hemisphere. In addition gyres often become ocean currents within larger, cooler, ocean movements, and can have an affect on regional or local weather systems, which in turn can affect the global climate. See SSTs, below.

I   Intertropical Convergence Zone (IGZ):   Low-latitude zones of convergence between air masses coming from either hemisphere of the boundary between the northern and southerly trade winds. It is connected with the westerly winds of the Pacific that bring on the monsoons, and the other meandering wind systems such as the *jet* *streams* which can have repercussions on the world's weather systems.

I Isostacy: The rise and fall of sea levels in relationship to similar or countervailing movements in the land.

L Latent heat transfer: The way the surface evaporation of water vapour from differentially heated regions of the world cools the Earth by convection processes. Humidity, atmospheric pressure, wind speeds and direction are part of this scenario, and arise from warm tropical waters dispersing thermal energy into the northern and southern hemispheres.

L Limit Cycle: Term used in chaos theory to describe the natural limiting factors beyond which catastrophic events occur. The biosphere is said to have built-in limit cycles to prevent a runaway greenhouse effect occurring. Chaos theorists say all dynamical systems seem to have a *strange attractor* which ensures that trajectories and trends keep within the limit cycle.

L Little Ice Age: A prolonged cool period occurring in the northern hemisphere in the 17th century, although some writers suggest it lasted from 1400 to about 1800. Winter temperatures were frequently up to 3C lower than they are today.

N Numerical Weather Prediction (NWP): Weather forecasts in numerical form as derived from the GCMs (see above). Based upon the prewar theories of Lewis Richardson, who first thought up the idea of assigning numerical variables to a checkerboard grid over a map of the Earth to calculate how the weather above the grid would respond to numerical changes.

P Persistence error: The percentage error made by weather-forecasters who predict that tomorrow's weather will be the same as today's which naturally increases as the forecast is extended over time. Persistence error is closely related to zero-change assumptions, where the monthly and seasonal weather is not expected to deviate from the average range of data.

P Polar-orbiting: An instrument-carrying satellite that orbits the Earth in a regular north-to-south pole configuration.

S Sea Surface Temperatures (SSTs): Sea current and gyres which variably show warmer and cooler temperatures, sometimes acting as 'rivers' within the oceans, some ephemeral but some having more significance, which oceanographers monitor to help them understand the relationship between ocean temperatures and climate.

S 'Sinks': Reservoirs of carbon within the Earth's biosphere or ecosystem; the degree to which these 'sinks' take in excess amounts of carbon dioxide and the way they determine the amount of CO^2 that remains in the atmosphere.

S '...sphere': The various Earth 'spheres' which scientists refer to represent the globe itself, with its concentric onion-like layers of rock, vegetation, atmosphere, oceans and ice. Many of them overlap. 'Ecosphere' and 'ecosytem' are similar to 'biosphere', and refer to the ecology or biological characteristics of the Earth. The biosphere is part of the Earth's environment in which living organisms are found, with which they interact to produce a self-contained system. The 'ecosphere' emphasizes the inter-connectedness of the biotic (the biological) and the abiotic (the non-biological) parts of the Earth. The abiotic more specifically refers to the 'geosphere', the geological parts of the Earth. The 'hydrosphere' refers to the planet's oceans and surface water, and includes the rain-bearing atmosphere. (see above). The 'cryosphere' refers to the Earth's stock of ice.

S Sunspots: Areas of deep magnetic turbulence on the surface of the Sun, apparently occurring in cycles of 7 to 10 years, which are alleged to affect weather systems on Earth via electromagnetic processes.

SELECT BIBLIOGRAPHY

The Ages of Gaia, James Lovelock, Oxford University Press, 1988.

Atmosphere, Weather and Climate, Roger G Barry & Richard J Chorley, 5th ed., Methuen, 1987.

The Co-Evolution of Climate and Life, S Schneider and R Londer, Sierra Club Books, 1984.

Earth's Changing Climate: The Cosmic Connection, Antony Milne, Prism Press, 1989.

Global Biomass Burning: Atmospheric, Climatic and Biospheric Implications, MIT Press, 1992.

Global Warming: Apocalypse or Hot Air?, Roger Bate and Julian Morris, IEA Environment Unit, 1994.

The Heated Debate, Robert C Balling, Jr, Pacific Research Inst (US), 1992.

Hothouse Earth, John Gribbin, Bantam Press, 1990.

Ice Ages: Solving the Mystery, J Imbrie & K Imbrie, Macmillan, 1979.

Intergovernmental Panel on Climate Change: Special Report: Radiative Forcing of Climate Change, 1990, WMO/UNEP.

Searching for Certainty, John L Casti, Abacus, 1993.

Sound and Fury, Patrick J Michaels, Cato Inst, Washington, 1992.

Weather, Climate & Human Affairs, H H Lamb, Routledge, 1988.

INDEX